The Challenges and Strategies of Food Security under Global Change

The Challenges and Strategies of Food Security under Global Change

Guest Editor

Raquel P. F. Guiné

MDPI

Basel • Beijing • Wuhan • Barcelona • Belgrade • Novi Sad • Cluj • Manchester

Guest Editor
Raquel P. F. Guiné
CERNAS Research Centre
Polytechnic University of Viseu
Viseu
Portugal

Editorial Office
MDPI AG
Grosspeteranlage 5
4052 Basel, Switzerland

This is a reprint of the Special Issue, published open access by the journal *Foods* (ISSN 2304-8158), freely accessible at: https://www.mdpi.com/journal/foods/special_issues/Security_Global_Change.

For citation purposes, cite each article independently as indicated on the article page online and as indicated below:

Lastname, A.A.; Lastname, B.B. Article Title. *Journal Name* **Year**, *Volume Number*, Page Range.

ISBN 978-3-7258-7218-3 (Hbk)
ISBN 978-3-7258-7219-0 (PDF)
https://doi.org/10.3390/books978-3-7258-7219-0

Cover image courtesy of Raquel P. F. Guiné

Contents

About the Editor

Raquel P. F. Guiné

Raquel P. F. Guiné is a Full Professor in the Food Industry Department at Polytechnic University of Viseu. She has a PhD in Chemical Engineering, a specialization in Food Engineering, and a habilitation in Food Science.

She has been president of the scientific board, president of the board of representatives, director of the department, and director of bachelor's and master's courses. She has been the coordinator of the CERNAS-IPV research unit and co-coordinator of the CI&DETS research centre.

The scientific areas of interest include food science, food engineering, food processing, food chemistry, nutrition, diets, and food consumption.

Global project coordinator of international projects like the EATMOT (18 countries: Portugal [Coord.], Argentina, Brazil, Croatia, Cyprus, Egypt, Slovenia, United States, Greece, Netherlands, Hungary, Italy, Latvia, Lithuania, Macedonia, Poland, Portugal, Romania, Serbia) and the EISuFood (18 countries: Portugal [Coord.], Brazil, Cape Verde, Colombia, Croatia, Slovenia, Spain, Greece, Latvia, Lithuania, Morocco, Mexico, Nigeria, Poland, Romania, Serbia, Turkey).

Regularly performs evaluation of scientific projects for FCT—Foundation for Science and Technology (Portugal); the Swiss National Science Foundation (Switzerland); European Science Foundation (France), Croatian Science Foundation (Croatia); and NCN—National Science Centre (Poland), in this last case assuming different roles, including that of coordinator of the panel of international experts. Participation in the organization of 32 events/conferences and a member of the Scientific Committee of 60 congresses/Meetings.

Authorship of 27 books, 115 chapters, 401 scientific articles, and 394 conference proceedings/abstracts. Communications at scientific and knowledge transfer meetings: 40 lectures/plenaries/Keynotes by invitation, 255 oral presentations, and 200 posters.

foods

MDPI

Editorial

The Challenges and Strategies of Food Security under Global Change

Raquel P. F. Guiné

CERNAS-IPV Research Centre, Polytechnic University of Viseu, 3504-510 Viseu, Portugal; raquelguine@esav.ipv.pt

1. The Challenges of Food Security

Food insecurity corresponds to a deficit in households' access to appropriate food, either in quantity and/or quality, due to limited financial resources or other factors. A related concept of nutrition insecurity is also gaining relevance, and this refers to a lack of access to food that has the adequate nutritional value necessary for a good health status and well-being [1].

Currently, the world faces troubling challenges that greatly impact food/nutritional security. One of the most worrying trends is an increase in the world's population and the rising number of countries that are food importers owing to their inability to produce the food they need to feed their population. The United Nations World Population Prospects [2] estimates a rise in the world's population from about 8 billion in 2025 to 10.5 billion by 2100. An increase in the world's population brings the added challenge of feeding such a large number of people with the Earth's limited resources. This is only possible with more efficient food production, more sustainable use of natural resources, and the minimization and reutilization of food and agricultural waste [3].

Complying with efficient food production while minimizing environmental impacts is a major issue since the more environmentally friendly agricultural practices, such as organic farming and integrated protection, do not allow the production of food at the same rates as conventional intensive farming practices [4,5]. Nevertheless, organic farming has a decisive role in achieving sustainability in agriculture, meeting a number of the Sustainable Development Goals of the United Nations [6]. In what concerns the minimization of agricultural and food waste and its reutilization, these goals have been at the top of priorities of academics, farmers, industrials, and even consumers in attempts to minimize waste along the whole food supply chain and to find alternative ways to value foods that are discarded [7–10].

One other challenge is related to the world's food supply being strongly under pressure due to the combined effects of environmental degradation and climate change. Environmental degradation, such as sterile agricultural land, depleted water tables, and falling grain yields, threatens food security for hundreds of millions of people, especially in more vulnerable regions of the globe [3,11,12]. Climate change has direct impacts on agriculture. Climate-induced drought reduces crop production and lowers yields, while warmer temperatures lead to a higher incidence of plant, livestock, and fish diseases [13,14]. Climate change is dislocating production in some regions, and this trend is expected to increase as the century progresses. The deforestation of tropical rainforests to obtain arable soil is a pivotal issue that contributes to environmental degradation while trying to respond to a need to increase the production of higher amounts of food [15–17].

One other challenge is related to a trend of rising populism allied to nationalism, which has increased trade protectionism. This brings huge constraints to the commercialization of foods on a global scale [18,19]. Allied with this challenge, there is the trend for an increase in the number of weak and at-risk states which are poorly equipped to address their food insecurity and, therefore, face social and/or political instability.

2. Wars and Conflicts as Enhancers of Food Insecurity Risks

War zones and political or social conflicts are major drivers for food insecurity, not only regionally but also on a global scale. Taking the example of the Russia–Ukraine conflict, it was evident that it would have a major impact on agricultural production and trade [20]. While Ukrainian wheat, soybean, and maize production fell abruptly in 2022–2023, the global supply chain and food trade were significantly affected, causing scarcity and a consequent increase in the world's food prices of these commodities [20]. The wheat crisis could thus be alleviated by increasing production and eliminating trade restrictions [21].

At more localized scales, the consequences of wars and conflicts are devastating for the affected populations, like, for example, those derived from the Israel–Palestine conflict in 2024 [22]. Although the food security of Palestinian households in the Gaza Strip was not directly impacted by the 2014 Gaza conflict due to some resilience and adaptive capacity [23], it is also a fact that these diminish with time and with increasing intensity of the conflict, as has happened in 2024 [22].

3. Prospects and Future Challenges

Scientists and politicians have to come together to examine global food security challenges and find the correct strategies to mitigate the risks. Resilient systems must be built to better face the adverse consequences of today's still inappropriate food systems. It is a fact that much has been done so far to achieve better efficacy in food production, utilization of natural resources and minimization of food and agricultural waste, but much more needs still to be achieved for a balanced utilization of resources while producing food able to satiate the increasing world population.

Success can only be achieved through research, implementation, and collaboration in a global strategy to fight a common problem. Individualism, nationalism, and politics tend to encourage too much self-focus on one's own success and are enemies of global perspectives and collaborative approaches.

Food production, food trade, climate change, environmental degradation, food supply chains and distribution networks, waste management, governance, and food policies are all issues that will have a decisive role in the future of food security/insecurity.

Funding: This research was funded by the FCT—Foundation for Science and Technology (Portugal) through projects Ref. UIDB/00681/2020 and UIDP/00681/2020. The APC was funded by FCT through projects Ref. UIDB/00681/2020 and UIDP/00681/2020.

Acknowledgments: This work was supported by the FCT—Foundation for Science and Technology, (Portugal). Furthermore, we would like to thank the FCT/CERNAS Research Center (projects doi:10.54499/UIDB/00681/2020 and doi:10.54499/UIDB/00681/2020) and the Polytechnic University of Viseu for their support.

Conflicts of Interest: The authors declare no conflicts of interest.

References

1. Scott, J.J.; de Bruin, W.B.; Rabinovich, L.; Frazzini, A.; de la Haye, K. Understanding Food Insecurity in Los Angeles County during the COVID-19 Pandemic and Its Aftermath: A Qualitative Interview Study. *Appetite* **2024**, *198*, 107323. [CrossRef] [PubMed]
2. UN World Population Prospects—Population Division—United Nations. Available online: https://population.un.org/wpp/ (accessed on 17 June 2024).
3. Manzoor, S.; Fayaz, U.; Dar, A.H.; Dash, K.K.; Shams, R.; Bashir, I.; Pandey, V.K.; Abdi, G. Sustainable Development Goals through Reducing Food Loss and Food Waste: A Comprehensive Review. *Future Foods* **2024**, *9*, 100362. [CrossRef]
4. Munné-Bosch, S.; Bermejo, N.F. Fruit Quality in Organic and Conventional Farming: Advantages and Limitations. *Trends Plant Sci.* **2024**. [CrossRef] [PubMed]
5. Wittwer, R.A.; Klaus, V.H.; Oliveira, E.M.; Sun, Q.; Liu, Y.; Gilgen, A.K.; Buchmann, N.; van der Heijden, M.G.A. Limited Capability of Organic Farming and Conservation Tillage to Enhance Agroecosystem Resilience to Severe Drought. *Agric. Syst.* **2023**, *211*, 103721. [CrossRef]
6. Gamage, A.; Gangahagedara, R.; Gamage, J.; Jayasinghe, N.; Kodikara, N.; Suraweera, P.; Merah, O. Role of Organic Farming for Achieving Sustainability in Agriculture. *Farming Syst.* **2023**, *1*, 100005. [CrossRef]

7. Rifna, E.J.; Dwivedi, M.; Seth, D.; Pradhan, R.C.; Sarangi, P.K.; Tiwari, B.K. Transforming the Potential of Renewable Food Waste Biomass towards Food Security and Supply Sustainability. *Sustain. Chem. Pharm.* **2024**, *38*, 101515. [CrossRef]
8. Fami, H.S.; Aramyan, L.H.; Sijtsema, S.J.; Alambaigi, A. The Relationship between Household Food Waste and Food Security in Tehran City: The Role of Urban Women in Household Management. *Ind. Mark. Manag.* **2021**, *97*, 71–83. [CrossRef]
9. Irani, Z.; Sharif, A.M.; Lee, H.; Aktas, E.; Topaloğlu, Z.; van't Wout, T.; Huda, S. Managing Food Security through Food Waste and Loss: Small Data to Big Data. *Comput. Oper. Res.* **2018**, *98*, 367–383. [CrossRef]
10. Knorr, D.; Augustin, M.A. From Kitchen Scraps to Delicacies to Food Waste. *Sustain. Food Technol.* **2024**, *2*, 652–666. [CrossRef]
11. Ashraf, J.; Javed, A. Food Security and Environmental Degradation: Do Institutional Quality and Human Capital Make a Difference? *J. Environ. Manag.* **2023**, *331*, 117330. [CrossRef] [PubMed]
12. Nguyen, T.T.; Grote, U.; Neubacher, F.; Rahut, D.B.; Do, M.H.; Paudel, G.P. Security Risks from Climate Change and Environmental Degradation: Implications for Sustainable Land Use Transformation in the Global South. *Curr. Opin. Environ. Sustain.* **2023**, *63*, 101322. [CrossRef]
13. Newbery, F.; Qi, A.; Fitt, B.D. Modelling Impacts of Climate Change on Arable Crop Diseases: Progress, Challenges and Applications. *Curr. Opin. Plant Biol.* **2016**, *32*, 101–109. [CrossRef]
14. Launay, M.; Caubel, J.; Bourgeois, G.; Huard, F.; de Cortazar-Atauri, I.G.; Bancal, M.-O.; Brisson, N. Climatic Indicators for Crop Infection Risk: Application to Climate Change Impacts on Five Major Foliar Fungal Diseases in Northern France. *Agric. Ecosyst. Environ.* **2014**, *197*, 147–158. [CrossRef]
15. Ghazoul, J.; Medina, L. Deforestation and Land Clearing. In *Encyclopedia of Biodiversity*, 3rd ed.; Scheiner, S.M., Ed.; Academic Press: Cambridge, MA, USA, 2024; pp. 605–616, ISBN 978-0-323-98434-8.
16. Cabral, B.F.; Yanai, A.M.; de Alencastro Graça, P.M.L.; Escada, M.I.S.; de Almeida, C.M.; Fearnside, P.M. Amazon Deforestation: A Dangerous Future Indicated by Patterns and Trajectories in a Hotspot of Forest Destruction in Brazil. *J. Environ. Manag.* **2024**, *354*, 120354. [CrossRef] [PubMed]
17. Gray, E. Tropical Forests and Water Security in South America: Deforestation Trends, Drivers, and Solutions for Water Suppliers and Regulators. In *Imperiled: The Encyclopedia of Conservation*; DellaSala, D.A., Goldstein, M.I., Eds.; Elsevier: Amsterdam, The Netherlands, 2022; pp. 145–156, ISBN 978-0-12-821139-7.
18. Hung, P.-Y. Food Nationalism beyond Tradition: Bubble Tea and the Politics of Cross-Border Mobility between Taiwan and Vietnam. *Asian J. Soc. Sci.* **2022**, *50*, 222–228. [CrossRef]
19. Dupraz, C.L.; Postolle, A. Food Sovereignty and Agricultural Trade Policy Commitments: How Much Leeway Do West African Nations Have? *Food Policy* **2013**, *38*, 115–125. [CrossRef]
20. Nasir, M.A.; Nugroho, A.D.; Lakner, Z. Impact of the Russian–Ukrainian Conflict on Global Food Crops. *Foods* **2022**, *11*, 2979. [CrossRef] [PubMed]
21. Lin, F.; Li, X.; Jia, N.; Feng, F.; Huang, H.; Huang, J.; Fan, S.; Ciais, P.; Song, X.-P. The Impact of Russia-Ukraine Conflict on Global Food Security. *Glob. Food Secur.* **2023**, *36*, 100661. [CrossRef]
22. Kimhi, A. Food Security in Israel: Challenges and Policies. *Foods* **2024**, *13*, 187. [CrossRef] [PubMed]
23. Brück, T.; d'Errico, M.; Pietrelli, R. The Effects of Violent Conflict on Household Resilience and Food Security: Evidence from the 2014 Gaza Conflict. *World Dev.* **2019**, *119*, 203–223. [CrossRef]

foods

MDPI

Article

Effects of Soaking and Germination Treatments on the Nutritional, Anti-Nutritional, and Bioactive Characteristics of Adzuki Beans (*Vigna angularis* L.) and Lima Beans (*Phaseolus lunatus* L.)

Qurat Ul Eain Hyder Rizvi [1], Raquel P. F. Guiné [2,*], Naseer Ahmed [1], Mohd Aaqib Sheikh [1,3], Paras Sharma [4], Imran Sheikh [5], Ajar Nath Yadav [5] and Krishan Kumar [1,6,*]

1 Department of Food Technology, Dr. Khem Singh Gill Akal College of Agriculture, Eternal University, Baru Sahib, Sirmour 173101, India; syeda15994@gmail.com (Q.U.E.H.R.); naseerfst@gmail.com (N.A.); sheikhaaqib111@gmail.com (M.A.S.)
2 CERNAS Research Centre, Polytechnic University of Viseu, 3504-510 Viseu, Portugal
3 Department of Food Process Engineering, National Institute of Technology, Rourkela 769005, India
4 Department of Food Technology, Mizoram University, Aizawl 796004, India; parassharma19791@gmail.com
5 Department of Biotechnology, Dr. Khem Singh Gill Akal College of Agriculture, Eternal University, Baru Sahib, Sirmour 173101, India; imran@eternaluniversity.edu.in (I.S.); ajar@eternaluniversity.edu.in (A.N.Y.)
6 Department of Food Technology, Rajiv Gandhi University, Doimukh 791112, India
* Correspondence: raquelguine@esav.ipv.pt (R.P.F.G.); krishan.kumar@rgu.ac.in (K.K.)

Abstract: Lima beans (*Phaseolus lunatus*) and adzuki beans (*Vigna angularis*) are some of the most nutritious underutilized pulses that are significant in being used as basic ingredients for the preparation of various food products. The present study aimed to determine the impact of soaking and germination on nutritional and bioactive components, in vitro protein digestibility, reducing power, metal chelating capacity, antioxidant activity, and anti-nutritional components of lima and adzuki beans. The findings showed that during the germination treatment, the in vitro protein digestibility of lima and adzuki beans increased by 14.75 and 10.98%, respectively. There was an increase in the antioxidant activity of lima beans by 33.48% and adzuki beans by 71.14% after 72 h of germination, respectively. The reducing power assay of lima and adzuki beans indicated an increase of 49.52 and 36.42%, respectively, during germination. Similarly, the flavonoid and metal chelating activity increased in lima and adzuki beans after 72 h of germination. In contrast, the anti-nutrients, such as phytic acid, tannin content, and trypsin inhibitor activity, decreased significantly $p < 0.05$ after 72 h of germination. These results are encouraging and allow for utilizing the flour obtained from the germinated beans in functional bakery products, which can contribute to eradicating protein deficiency among some population groups. At the same time, promoting soaking and germination of the beans as a way to enhance the nutritional quality and reduce anti-nutrients can contribute to the interest in these underutilized pulses. They could be seen as an additional tool to improve food security.

Keywords: antioxidant activity; anti-nutrients; bioactive components; germination; nutritional compounds; soaking

1. Introduction

The most serious problem the developing countries are facing is protein-energy malnutrition. The developing world is facing a major challenge in providing nutritious, safe, and wholesome food for undernourished and poor populations. The search for reliable and affordable sources of protein derived from plants has been prompted by the unpredictability of supply, high costs, and scarcity of foods rich in protein from animals in underdeveloped and developing nations. Some of the underutilized and wild legumes are found to contain rich nutraceutical value. Furthermore, the presence of various anti-nutritional components restricts their utilization in food products, and these can be eliminated by employing certain processing methods like soaking and germination [1].

Among legumes, lima beans (*Phaseolus lunatus*) and adzuki beans (*Vigna angularis*) are considered underutilized pulses in India. Lima beans and adzuki beans are herbaceous plants belonging to the family Fabaceae. Lima beans are also referred to as faba beans, sugar beans, and butter beans. Lima beans are a great source of carbohydrates, dietary fibers, and proteins but contain lesser amounts of fat. It is rich in riboflavin, vitamin B6, thiamin, and niacin, which act as coenzymes for the oxidation of carbohydrates, proteins, and fats [2]. Lima beans can be utilized as whole beans and can also be converted into flour, which can be incorporated into conventional flour for better nutrient content. Lima beans are still underutilized pulses despite having a great nutritional profile. Matured lima bean seeds possess various benefits to human health. Despite various nutritional benefits, both lima and adzuki beans contain various anti-nutrients like phytic acid, tannins, trypsin inhibitors, oxalate, haematoglutinins, and cyanides, which interfere with the utilization and absorption of numerous micronutrients, thereby decreasing the nutritive value and protein digestibility of foods [3].

Adzuki beans are a traditional crop that plays a great role in the sustainability of agriculture and food protein supply. It is known for its traditional therapeutic uses. Adzuki beans have raised global interest not only because of their nutritional properties, health benefits, and processing qualities but also due to their gluten-free nature. Various studies reveal the highest nutrient profile of adzuki and have considered it to be a functional food for disease prevention and health promotion [4]. Adzuki beans are also known for their therapeutic benefits as they are anti-inflammatory, anti-diabetic, and antioxidant. They are also used as herbal medicine that controls weight, maintains health, and acts as a functional food material [5]. Eliminating or reducing these anti-nutrients to a safe level enhances the nutritive value of the food and also ensures the effective use of these legumes for human consumption.

Conventionally, grains are hydrated in water by soaking [6], which has been shown to be effective in both reducing and eliminating the anti-nutrients found in cereals and pulses [7]. Trypsin inhibitors, phytic acid, and other anti-nutrients that are either fully or partially soluble in water can be reduced in concentration by soaking cereals and legumes for 12 to 18 h, according to multiple studies [6,8]. Rizvi et al. [9] reported that the soaking and germination techniques were found to decrease the anti-nutritional components such as phytic acid and tannin contents in processed pigeon pea grains. There was an increase in the antioxidant activity and the total phenolic contents of germinated pigeon pea grains.

Various investigations have recommended that anti-nutritional factors in beans or legumes can be eliminated or decreased by many processing techniques like germination, fermentation, boiling, cooking, and soaking. Germination is a commonly used conventional technique that increases some anti-nutritional components and improves bioactive components while improving nutrient digestibility [9]. During the germination and soaking process, certain alterations could occur within the seed, and these vary depending on the variety of the seed and the germination condition. Germination is known to be the best technique for reducing the anti-nutritional compounds of pulses [10].

Chauhan et al. [11] studied the effect of soaking, germination, fermentation, and roasting on the nutrients, anti-nutrients, and bioactive components of black soybean. The results revealed that the phenolic contents augmented significantly ($p < 0.05$) in germination, fermentation, and roasting. The antioxidant activity of processed grains increased significantly ($p < 0.05$) during germination and fermentation. The anti-nutritional compounds, such as phytic acid and tannin contents, decreased significantly during processing treatments. The consumption of sprouts has increased because of convenience and complex physiological changes that increase the nutritional profile and reduce anti-nutrients that interfere with absorption.

In order to improve the nutritional quality of functional food products made from these underutilized pulses, the current research was designed to examine the impact of soaking and germination processing treatments on the nutritional, anti-nutritional, and bioactive potential of these pulses. This was performed in light of the therapeutic and nutritional benefits of these underutilized pulses as well as the significance of these processing treatments. Because processed grains have fewer anti-nutritional components, this study also supports the use of traditional processing methods to increase the nutritional value and bioavailability of micronutrients from underutilized pulses.

2. Materials and Methods

2.1. Materials

The lima beans (American white variety) and adzuki beans (HPU-51) used in the present study were procured from Bombay super seeds, Dist Rajkot (Gujrat), India, and CSK Palampur, Himachal Pradesh, respectively. In the current investigation, ultrapure-grade chemicals and reagents were employed. These were acquired from typical chemical suppliers, including Sigma-Aldrich, Bangalore, India; Hi-Media, Thane, India; Qualigens, Mumbai, india and Merck, Mumbai India. The 3 kg seeds of each of the lima and adzuki beans were procured, and 250 g seeds of each bean type were taken for processing treatments at different time intervals.

2.2. Methods

2.2.1. Soaking and Germination Process

The soaking and germination of lima and adzuki beans were carried out using the procedures outlined by Egli et al. [12]. After eliminating any foreign particles, distilled water was used to soak the grains in a 1:5 ratio. This process was performed for both lima and adzuki beans. The seeds were soaked at room temperature for 12 (S12) and 24 (S24) h, after which they were dried for 24 h at 40 °C in a hot air oven. After being placed in airtight pouches for additional analysis, the grains were kept cold at 4 °C.

After steeping, the seeds were drained and covered with the damp muslin cloth, placed in a seed germinator (Western Agro, Helix Technology), and removed at intervals of 24, 48, and 72 h. The germination was carried out for 0 (control), 24 (G24), 48 (G48), and 72 h (G72) in the absence of light at a temperature of 25 °C and 80% relative humidity. Following each germination time treatment, the seeds were dried in a hot air oven at 40 °C for 24 h and ground into fine flour with a laboratory flour mill (SANCO). After that, the flour made from germinated pulses was kept at 4 °C until further examination (Figure 1). The seeds of both the lima and adzuki beans at 0 h without any treatment were kept as a control for comparison with samples treated with soaking and germination treatment at different time intervals.

Figure 1. Soaking and germination treatments of lima and adzuki beans at different time intervals (RG—raw grains; S 24—soaking for 24 h; G24—germination for 24 h; G48—germination for 48 h; G72—germination for 72 h).

2.2.2. Physical and Functional Characteristics

The evaluation of the physical characteristics of raw, soaked, and germinated adzuki and lima beans was carried out using standard procedures. Physical parameters like the thousand-grain weight (TGW) were determined by measuring the weight of a thousand grains of adzuki and lima beans and expressed in grams [13]. The length, breadth, and thickness were determined with the help of the Vernier caliper. Bulk density (BD) was evaluated according to the method of Huang et al. [14], and tap density (TD) was evaluated as per the method given by Jones et al. [15]. The water absorption capacity (WAC) was determined as per the methodology described by Sosulski [16], and the oil absorption capacity (OAC) was determined by Kaur et al. [17] after converting grains to flour. The water solubility index (WSI) was estimated using the method of Stojceska et al. [18]. The color characteristics, such as L^*, a^*, and b^* values, were estimated with the help of a chromameter (CR-400, Konika Minolta, Tokyo, Japan).

2.2.3. Chemical Characteristics

The hot air oven method, as described by AOAC [19], was used to estimate the moisture content (%) of the grains. The crude fiber was estimated using Fibroplus FBS 08P (Pelican Inc., Torrance, CA, USA), crude fat was estimated using Soxoplus SPS 06 AS (Pelican Inc.), and crude proteins were estimated using Kjelodist CAS VA (Pelican Inc.). The ash content was calculated using the procedures outlined by Ranganna [20]. The analysis of mineral components was conducted using an Atomic Absorption Spectrometer (AA240FS, Agilent Technology, CA, USA) to determine the presence of iron, zinc, manganese, and copper [19]. By subtracting the measured amounts of moisture, crude protein, ash, crude fat, and crude fiber from 100, the total carbohydrate contents were determined. The factors of 4.0, 9.1, and 4.2 kcal/g for crude protein (Nx6.25), fats, and carbohydrates, respectively, were used to estimate the calorific value (Kcal/100 g) [21]. The methodology, as defined by Shastry and John [22], was used to assess the in vitro protein digestibility.

Flavonoid content was determined using the method described by Lahlou et al. [23]. The 2.5 g sample was added to a 10 mL distilled water solution and kept overnight at room temperature (37 °C). The mixture was centrifuged at 8000 rpm for 15 min, and the clear supernatant was obtained. One mL of 2% $AlCl_3$ was dissolved in 80% methanol, mixed with the 1 mL of extract solution, and incubated for 1 h at 37 °C. Absorbance was measured at 415 nm using a spectrophotometer. Total flavonoid contents were determined using a standard curve of rutin. The results were expressed as mg/100 g rutin equivalent (RE) of the sample. The antioxidant activity (%) was calculated using the method described by Bouaziz et al. [24] by measuring the 2,2-diphenyl-1-picrylhydrazyl (DPPH) radical scavenging capacity.

The metal chelating activity and reducing capacity were calculated using the procedure as defined by Sharma and Gujral [25]. In the case of metal chelating activity, 0.4 mL $FeSO_4$ and 0.8 mL ferrozine were subsequently added into 0.4 mL of plant extract. The mixture was incubated at room temperature for 10 min, and the absorbance of the mixture was measured at 562 nm. The metal chelating activity was calculated using the following formula:

$$\text{Metal chelating activity}(\%) = \frac{\text{Absorbance of control} - \text{absorbance of sample}}{\text{Absorbance of control}} \times 100$$

For estimation of reducing capacity, the 2.5 mL of phosphate buffer, 2.5 mL of potassium ferricyanide, and 2.5 mL of TCA were added to 1 mL of sample extract and incubated at 50 °C for 20 min. The above mixture was centrifuged at 3000 rpm for 10 min. An amount of 0.5 mL of ferric chloride was added to the above mixture, and the absorbance was measured at 700 nm using a spectrophotometer. A blank was prepared without adding extract. Ascorbic acid at various concentrations (10–100 µg/mL) was used as standard. The increased absorbance of the reaction mixture indicates an increase in reducing capacity (%).

The tannin contents were estimated using the method of Saxena et al. [26], while the trypsin inhibitor activity (TI) was ascertained using the method of Kakade et al. [27]. The phytic acid was assessed using the methodology of Gao et al. [28] with minor modifications.

The total phenolic content was determined using a Folin–Ciocalteu reagent, followed by a method of Ainsworth and Gillespie [29] with minor modifications. A volume of

0.5 mL of the plant extract was mixed with 2 mL of the Folin–Ciocalteu reagent (diluted 1:10 with deionized water) and neutralized with 4 mL of sodium carbonate solution (7.5%, *w*/*v*). Gallic acid was used as a reference standard for plotting the calibration curve. The color developed by the incubation of the reaction mixture for 30 min was measured at an absorbance of 765 nm using a visible spectrophotometer (711-S NV). The total phenolic contents were assessed from the linear equation of a standard curve prepared with Gallic acid. The content of total phenolic compounds expressed as mg/100 g gallic acid equivalent (GAE) of dry extract.

2.2.4. FT-IR (Fourier Transform Infrared Spectroscopy) Analysis

The FTIR spectroscopic analysis was carried out using a Fourier-transform spectrophotometer (PerkinElmer, Waltham, MA, USA) in the range of 600–4000 cm^{-1}.

2.2.5. Statistical Analysis

All experiments were performed in triplicate, and the data collected during the study were subjected to one-way analysis of variance (ANOVA) using the IBM SPSS Statistics 26 software. Based on Duncan's Multiple Range test post hoc analysis, differences were deemed significant at the $p < 0.05$ level, and values in the tables were expressed as mean ± standard deviation. Principal component analysis (PCA) of all analytical variables was performed using Minitab 16 statistical software.

3. Results and Discussion

3.1. Physical Properties

The moisture content of lima and adzuki beans was 7.29 and 12.32%, respectively. Farinde et al. [30] reported 7.50% moisture content in lima beans. The initial higher moisture content of grains during storage decreases the germination capacity of grains [31]. The length, width, thickness, and thousand-grain weight (TGW) observed in lima beans were 13.26 mm, 9.87 mm, 9.16 mm, and 699.65 g, respectively. Purwanti and Fauzi [32] reported 15.7 and 10.8 mm of length and width in lima beans, and the TGW observed was 775 g. Similarly, the length, width, thickness, and TGW observed in adzuki beans were 4.32 mm, 2.73 mm, 2.36 mm, and 61.27 g, respectively. Agarwal and Chauhan [4] reported 4.2 mm and 3 mm of length and width, respectively, in adzuki beans. Wu et al. [33] found 58.27 g of TGW in adzuki beans, whereas Yadav et al. [26] reported 74.87 g of TGW in adzuki beans (Table 1).

Table 1. Physical and functional characteristics of lima and adzuki beans.

Parameters	Lima Beans	Adzuki Beans
Moisture (%)	7.29 ± 0.07 ^{b}	12.32 ± 0.08 ^{a}
Length (mm)	13.26 ± 0.16 ^{a}	4.32 ± 0.03 ^{b}
Width (mm)	9.87 ± 0.12 ^{a}	2.73 ± 0.32 ^{b}
Thickness (mm)	9.16 ± 0.06 ^{a}	2.36 ± 0.06 ^{b}
1000 Grain Wt. (g)	699.65 ± 0.64 ^{a}	61.27 ± 1.05 ^{b}
Bulk density (g/cm^3)	0.93 ± 0.06 ^{a}	0.86 ± 0.01 ^{a}
Tap density (g/cm^3)	1.26 ± 0.05 ^{a}	1.16 ± 0.03 ^{a}
Water Absorption Capacity (mL/g)	2.21 ± 0.05 ^{a}	2.89 ± 0.11 ^{a}
Oil Absorption Capacity (mL/g)	1.66 ± 0.05 ^{a}	1.81 ± 0.11 ^{a}
Water Solubility Index (%)	20.23 ± 0.10 ^{b}	23.23 ± 0.09 ^{a}
Swelling Capacity (%)	82.70 ± 0.46 ^{a}	70.06 ± 0.03 ^{b}
L value	82.34 ± 0.07 ^{a}	44.14 ± 0.80 ^{b}
a^* value	−0.56 ± 0.07 ^{b}	4.33 ± 0.10 ^{a}
b^* value	8.07 ± 0.03 ^{a}	0.86 ± 0.04 ^{b}

Values are expressed as mean ± SD. Duncan's LSD post hoc analysis at $p < 0.05$ indicates that values within rows sharing the same letter are not significantly different.

3.2. Functional Properties

The bulk density and tap density observed in lima beans were 0.93 and 1.26 g/cm^3, and in adzuki beans, the values were observed as 0.86 and 1.16 I g/cm^3, respectively (Table 1). Yadav et al. [34] observed 0.76 g/mL of bulk density in adzuki beans. The water absorption capacity (WAC) and oil absorption capacity (OAC) were observed as 2.21 and 1.66 mL/g, respectively, in lima beans and 2.89 and 1.81 mL/g, respectively, in adzuki beans. The study carried out by Siddiq et al. [35] reported 2.25 and 1.52 g/g of WAC and OAC, respectively, in red kidney beans. The water solubility index (WSI) of lima and adzuki beans was observed to be 20.23 and 23.23%. Yellavila et al. [3] reported 21.01% WSI in lima beans. The swelling capacities observed in lima and adzuki beans were 82.70 and 70.06%, respectively. The color characteristics such as L^*, a^*, and b^* values observed in lima beans were 82.34, −0.56, 8.07, and that in adzuki beans were 44.14, 4.33, and 0.86, respectively. Woo et al. [36] observed 68.97% of swelling capacity in adzuki beans and reported 43.66, 5.07, and 0.89 L^*, a^*, and b^* values in adzuki beans (Table 1).

3.3. Principal Component Analysis

The results of principal component analysis (PCA) of all analytical variables are presented in Figure 2 and Table 2.

Table 2. Principal component analysis and loading of first four components.

Factor Number	Lima Beans				Adzuki Beans			
	1	2	3	4	1	2	3	4
Initial eigenvalues	11.245	6.06	0.000	0.000	11.344	4.66	0.000	0.000
% of variance	0.937	0.050	0.006	0.000	0.945	0.039	0.000	0.000
Cumulative %	93.70	98.80	100	100	94.50	98.40	100	100
Factor Loadings								
Moisture content (%)	−0.291	0.166	−0.510	0.503	0.288	0.310	0.297	0.097
Fat content (%)	0.275	**0.452**	−0.576	−0.191	−0.291	0.053	−0.384	**0.632**
Fiber content (%)	−0.282	−0.395	−0.089	0.416	**0.294**	0.078	−0.301	−0.034
Ash content (%)	−0.296	−0.130	0.029	−0.274	0.253	−0.753	0.247	0.069
Protein content (%)	−0.283	0.356	**0.446**	0.290	0.286	**0.363**	−0.225	−0.095
Carbohydrate (%)	0.287	−0.341	0.058	0.294	0.289	−0.241	−0.283	0.420
Calorific value (Kcal/100 g)	**0.296**	0.081	0.043	**0.423**	−0.296	0.066	0.000	0.086
In vitro protein digestibility (%)	−0.292	0.227	0.244	−0.148	0.295	0.088	0.220	−0.100
Phytic acid (mg/100 g)	0.291	−0.274	0.097	−0.057	−0.292	−0.150	−0.302	−0.574
Tannin contents (mg/100 g)	0.278	0.446	0.342	0.204	−0.293	−0.061	**0.451**	0.157
Trypsin-inhibitor activity (TIU/100 g)	0.296	−0.136	0.086	−0.106	−0.291	0.238	0.332	0.138
Antioxidant activity (% DPPH scavenging capacity)	−0.296	−0.037	−0.044	−0.188	0.293	0.207	0.170	0.095

1—PC1; 2—PC2; 3—PC3; 4—PC4; PC—principal component. The values in bold indicates the dominating factor in each principal component.

Using the Kaiser criterion, the principal components (PCs), with eigenvalues greater than one (11.245 in lima beans and 11.344 and 4.66 in adzuki beans), were selected. The principal components from 1 to 4 contributed to more than 100 percent of the difference in the selected samples. The PC1, PC2, PC3, and PC4 reported 93.70, 98.80, 100, and 100% for lima beans and 94.50, 98.40, 100, and 100 percent of the variance for adzuki beans, respectively. Out of the four main components reported, the PC1 reported 93.70% of the variance for lima beans and 94.50% of the variance for adzuki beans, which was mostly dominated by calorific value in the case of lima beans and crude fiber in the case of adzuki beans shown in bold in Table 2. The second component was dominated by fat in the case of lima beans and protein in the case of adzuki beans, while protein content and tannin dominated the third component for lima beans and adzuki beans, respectively, and the last component was dominated by calorific value and fat content in case of lima and adzuki beans, respectively.

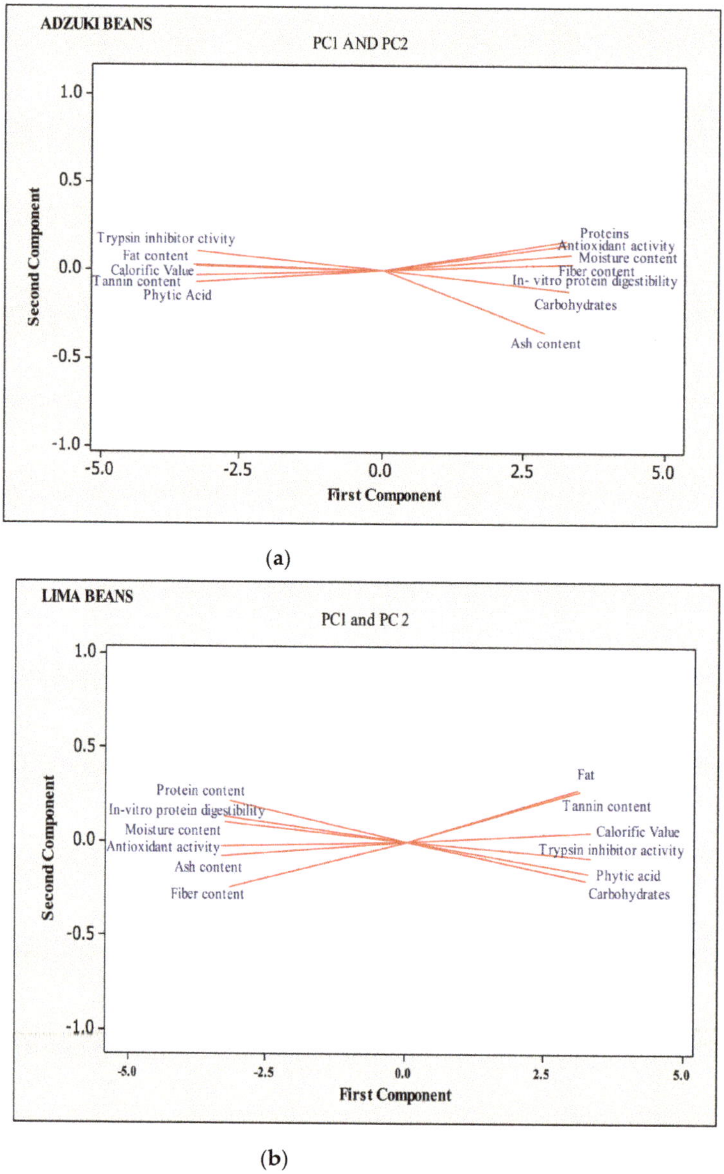

Figure 2. Projections of the variables on the factor plane for adzuki (**a**) and lima beans (**b**).

Therefore, in the current results, most of the proximate components were important variables that had the ability to discriminate with other physico-chemical properties. Figure 2 shows that all samples on the upper side of PC1 were related to the calorific value, while other samples placed on its upper left were related to proximate and tannins. Thus, the proximate composition could be used to differentiate the samples from each other.

3.4. Chemical Characteristics of Lima and Adzuki Beans

The fat, protein, ash, and fiber content observed were 1.47, 18.59, 3.78, and 6.55%, respectively, in lima beans and 3.16, 19.56, 3.29, and 5.79%, respectively, in adzuki beans (Tables 3 and 4). Farinde et al. [30] reported 7.50, 1.22, 22.24, 4.68, and 6.85% of moisture, fat, protein, ash, and fiber contents, respectively, in lima beans, and Yu-Wei and Wang [37] reported 3.1, 22.65, 1.87, and 3.41% of ash, protein, fat, and fiber content in raw adzuki beans.

The in vitro protein digestibility observed in lima and adzuki beans was 67.85 and 70.56%. The results observed by Alonso et al. [38] showed 68.1% in vitro protein digestibility in kidney beans. The phenolic and flavonoid content observed in raw lima beans was 110.43 and 20.16 mg/100 g, and the same in adzuki beans were observed as 22.78 and 43.38 mg QE/100 g, respectively. Khang et al. [39] reported 12.21 mg/100 g of phenolic content in raw adzuki beans. Meanwhile, Sombié et al. [40] reported 23.95 mg QE/100 g of flavonoid content in raw cowpeas seeds. The tannin content found in raw lima and adzuki beans was 10.70 and 6.36 mg/100 g, respectively. Similar results were found by Farinde et al. [30], who reported 9.80 mg/100 g of tannin content in raw lima beans. Phytic acid content found in lima and adzuki beans was 20.48 and 7.12 mg/100 g. Similar results were found by Adeparusi [41], who reported 19.0 mg/100 g of phytic acid in raw lima beans.

The antioxidant activity and reducing activity found in raw lima and adzuki beans were found to be 35.48, 30.55, and 9.74, 32.77%, respectively. Similar results were found by Yu-Wei and Wang [37], who reported 36.24 and 10.82% of antioxidant activity in faba beans and adzuki beans, respectively. Similarly, Granito et al. [42] found 49.78% of antioxidant activity in lima beans. The copper, iron, zinc, and manganese content found in raw lima beans were 4.17, 4.04, 2.45, and 2.36 ppm, and that in adzuki beans were 1.22, 5.16, 4.49, and 1.57 ppm, respectively. Similar results were observed by Jayalaxmi et al. [43], who observed 4.30, 5.40, 2.64, and 2.63 mg/100 g of copper, iron, zinc, and manganese content in raw lima beans.

Table 3. Changes in nutritional characteristics of lima beans during soaking and germination treatments.

Parameters	Time (h)					
	Soaking			Germination		
	0	12	24	24	48	72
Moisture content (%)	7.29 ± 0.07 d	7.31 ± 0.01 d	7.38 ± 0.01 c	7.43 ± 0.01 c	7.57 ± 0.02 b	7.74 ± 0.05 a
Protein content (%)	18.59 ± 0.61 e	18.24 ± 0.01 e	19.15 ± 0.01 d	20.07 ± 0.05 c	20.94 ± 0.04 b	21.92 ± 0.04 a
Fat content (%)	1.47 ± 0.25 a	1.36 ± 0.30 a	1.28 ± 0.11 ab	1.23 ± 0.08 ab	1.19 ± 0.08 ab	1.03 ± 0.01 b
Ash content (%)	3.78 ± 0.06 b	4.26 ± 0.63 a b	4.29 ± 0.08 ab	4.34 ± 0.28 a	4.41 ± 0.06 a	4.49 ± 0.06 a
Fiber content (%)	6.55 ± 0.03 e	6.79 ± 0.04 d	6.88 ± 0.06 d	7.37 ± 0.15 c	7.66 ± 0.14 b	7.94 ± 0.05 a
Carbohydrate content (%)	62.03 ± 0.46 a	62.04 ± 0.61 b	61.02 ± 0.11 c	59.56 ± 0.43 d	58.21 ± 0.26 e	56.89 ± 0.05 f
Calorific value (kcal/100 g)	350.96 ± 1.48 a	345.86 ± 3.90 b	344.50 ± 0.72 bc	341.67 ± 1.12 cd	339.14 ± 0.46 d	335.95 ± 0.13 e
Phytic acid (mg/100 g)	20.48 ± 0.13 a	19.20 ± 0.06 b	18.07 ± 0.02 c	17.67 ± 0.42 c	14.58 ± 0.41 d	12.68 ± 0.42 e
Tannin content (mg/100 g)	10.70 ± 0.08 a	9.85 ± 0.05 b	9.05 ± 0.04 c	6.57 ± 0.12 d	5.49 ± 0.06 e	4.35 ± 0.12 f
Phenolic content (mgGAE/100)	110.43 ± 0.16 f	115.27 ± 0.04 e	120.59 ± 0.04 d	125.41 ± 0.07 c	130.32 ± 1.17 b	139.74 ± 0.12 a
Antioxidant activity (% DPPH scavenging capacity)	35.48 ± 0.16 f	36.70 ± 0.08 e	38.59 ± 0.07 d	40.28 ± 0.06 c	43.71 ± 0.02 b	47.36 ± 0.08 a
Trypsin inhibitor (TIU/100 g)	20.41 ± 0.36 f	14.40 ± 0.34 e	13.13 ± 0.12 d	7.68 ± 0.27 c	5.53 ± 0.01 b	3.22 ± 0.09 a
Flavonoid content (mgQE/100 g)	20.16 ± 0.05 f	21.49 ± 0.07 e	23.19 ± 0.10 d	28.63 ± 0.26 c	30.62 ± 0.30 b	33.63 ± 0.14 a
Reducing capacity (%)	30.55 ± 0.05 f	33.66 ± 0.09 e	35.19 ± 0.16 d	39.74 ± 0.16 c	42.21 ± 0.06 b	45.68 ± 0.12 a
Metal chelating activity (%)	35.51 ± 0.07 f	39.06 ± 0.71 e	42.88 ± 0.11 d	47.48 ± 0.16 c	51.80 ± 0.17 b	55.42 ± 0.07 a
In vitro protein digestibility (%)	67.85 ± 0.09 f	69.84 ± 0.09 e	70.80 ± 0.25 d	72.42 ± 0.09 c	74.65 ± 0.08 b	77.86 ± 0.05 a
Zinc (ppm)	2.45 ± 0.03 bcd	2.33 ± 0.02 cd	2.25 ± 0.04 d	2.57 ± 0.04 abc	2.66 ± 0.03 ab	2.76 ± 0.32 a
Iron (ppm)	4.04 ± 0.04 e	4.15 ± 0.03 d	4.25 ± 0.03 c	4.46 ± 0.09 b	4.52 ± 0.08 b	4.96 ± 0.03 a
Copper (ppm)	4.17 ± 0.02 d	4.05 ± 0.03 e	3.97 ± 0.03 f	4.25 ± 0.04 c	4.94 ± 0.03 b	5.27 ± 0.02 a
Manganese (ppm)	2.36 ± 0.03 f	2.16 ± 0.03 e	2.27 ± 0.02 d	3.17 ± 0.02 c	3.34 ± 0.03 b	3.55 ± 0.04 a

The values in the table are displayed as mean ± standard deviation; Duncan's LSD post hoc analysis at $p < 0.05$ indicates that values within rows sharing the same letter are not significantly different.

3.5. *Effect of Processing Treatments on Nutritional, Anti-Nutritional, and Bioactive Characteristics*

3.5.1. Nutritional Characteristics and In Vitro Protein Digestibility

The moisture content increased in lima beans from 7.29 (RG) to 7.74% (G72) and 12.32 (RG) to 14.57% (G72) in adzuki beans, respectively (Tables 3 and 4). The moisture content increased by 1.23 and 6.17% in lima beans after soaking and germination treatments, respectively. Similarly, it increased by 9.00 and 18.26% after soaking and germination treatments, respectively, in adzuki beans. Farinde et al. [36] reported a 4.66% rise in moisture content in germinated lima beans, whereas Mubarak [44] reported a 13.84% increase in moisture content in mung bean sprouts. The increase in the moisture content may be attributed to the absorption of water from the surroundings during soaking and germination of beans to commence the metabolic processes [45].

Table 4. Changes in nutritional characteristics of adzuki beans during soaking and germination treatments.

Parameters	Time (h)					
	Soaking			Germination		
	0	12	24	24	48	72
Moisture content (%)	12.32 ± 0.08 f	12.87 ± 0.02 e	13.43 ± 0.06 d	13.59 ± 0.12 c	13.92 ± 0.06 b	14.57 ± 0.03 a
Protein content (%)	19.56 ± 0.03 f	20.96 ± 0.04 e	21.26 ± 0.03 d	22.76 ± 0.03 c	22.97 ± 0.03 b	23.46 ± 0.02 a
Fat content (%)	3.16 ± 0.02 c	3.26 ± 0.03 b	3.35 ± 0.04 a	2.85 ± 0.03 d	2.54 ± 0.02 e	1.95 ± 0.03 f
Ash content (%)	3.29 ± 0.07 e	3.34 ± 0.04 e	3.41 ± 0.04 d	3.54 ± 0.03 c	3.75 ± 0.02 b	3.90 ± 0.06 a
Fiber content (%)	5.79 ± 0.09 d	5.80 ± 0.06 d	5.91 ± 0.08 d	6.19 ± 0.08 c	6.59 ± 0.08 b	7.47 ± 0.06 a
Carbohydrate content (%)	55.87 ± 0.17 a	53.77 ± 0.06 b	52.48 ± 0.08 c	51.24 ± 0.07 d	49.57 ± 0.09 e	49.32 ± 0.06 f
Calorific value (kcal/100 g)	341.69 ± 0.75 a	339.38 ± 0.33 b	335.94 ± 0.71 c	332.14 ± 0.45 d	323.22 ± 0.21 e	318.42 ± 0.35 f
In vitro protein digestibility	70.56 ± 0.23 f	72.16 ± 0.06 e	73.62 ± 0.16 d	75.89 ± 0.07 c	77.54 ± 0.05 b	78.31 ± 0.06 a
Phytic acid (mg/100 g)	7.12 ± 0.03 a	6.15 ± 0.04 b	5.34 ± 0.02 c	4.66 ± 0.03 d	3.98 ± 0.01 e	3.45 ± 0.03 f
Tannin content (mg/100 g)	6.36 ± 0.01 a	6.22 ± 0.01 a	6.09 ± 0.09 a	5.91 ± 0.04 a	4.20 ± 0.25 b	2.47 ± 0.53 c
Trypsin inhibitor (TIU/100 g)	35.23 ± 0.11 a	30.57 ± 0.04 b	25.21 ± 0.06 c	20.64 ± 0.10 d	15.54 ± 0.09 e	10.21 ± 0.06 f
Phenolic content (mg/g)	16.16 ± 1.30 e	18.06 ± 0.72 d	19.21 ± 0.59 cd	20.17 ± 0.33 bc	21.27 ± 1.28 ab	22.39 ± 0.72 a
Antioxidant Activity (% DPPH scavenging capacity)	9.74 ± 0.04 cd	9.92 ± 0.06 cd	10.04 ± 0.03 cd	12.24 ± 0.03 bc	14.98 ± 0.02 ab	16.67 ± 0.02 a
Reducing capacity (%)	28.77 ± 0.02 f	30.04 ± 0.03 e	32.95 ± 0.03 d	35.27 ± 0.02 c	37.65 ± 0.04 b	39.25 ± 0.02 a
Flavonoid content (mg QE/100 g)	25.38 ± 0.18 f	26.89 ± 0.05 e	28.29 ± 0.07 d	30.64 ± 0.17 c	32.34 ± 0.10 b	34.47 ± 0.31 a
Metal chelating activity (%)	20.59 ± 0.05 f	23.80 ± 0.14 e	26.60 ± 0.33 d	28.62 ± 0.12 c	31.52 ± 0.43 b	34.64 ± 0.11 a
Zinc (ppm)	4.49 ± 0.04 f	4.92 ± 0.05 e	5.23 ± 0.04 d	5.94 ± 0.05 c	6.39 ± 0.05 b	6.74 ± 0.04 a
Iron (ppm)	1.57 ± 0.02 e	1.61 ± 0.02 e	1.73 ± 0.05 d	1.85 ± 0.03 c	1.97 ± 0.03 b	2.22 ± 0.03 a
Copper (ppm)	1.22 ± 0.02 a	1.49 ± 0.35 a	1.35 ± 0.07 a	1.41 ± 0.11 a	1.45 ± 0.04 a	1.51 ± 0.02 a
Manganese (ppm)	5.16 ± 0.04 f	5.35 ± 0.04 e	5.51 ± 0.06 d	6.34 ± 0.11 c	6.82 ± 0.07 b	7.32 ± 0.03 a

The values in the table are displayed as mean ± standard deviation; Duncan's LSD post hoc analysis at $p < 0.05$ indicates that values within rows sharing the same letter are not significantly different.

The protein content in lima beans increased from 18.59 (RG) to 21.92% (G72 h) and from 19.56 (RG) to 23.46% (G72 h) in adzuki beans. The protein content in lima and adzuki beans increased by 3.01%, 17.19%, and 8.69%, 19.93%, respectively, after soaking and germination treatments. Jayalaxmi et al. [43] observed an increase of 4.47 and 7.72% in protein content, respectively, after soaking and germination treatments in lima beans. The soaking and germination increased the protein content in the lima beans due to the induction of hydrolytic enzymes, which increased the metabolic activity of protein synthesis in seeds. The reawakening of protein synthesis upon imbibition leads to the escalation in the protein content of sprouted seeds [45].

The fat content in lima and adzuki beans decreased from 1.47 (RG) to 1.03% (G72 h) and 3.16 (RG) to 1.95% (G72 h), respectively. The fat content decreased in lima and adzuki beans to 29.93 and 38.29%, respectively, after germination; the results obtained in the current study were similar to the findings of Jayalaxmi et al. [43], who found a 17.85% decrease in fat content after germination treatment. Lima beans are low in fat, and it is further decreased due to fat being utilized during metabolic activities in the germinated seeds [36]. The decrease in fat content during the germination process may also be due to the transformation of fatty acids into carbohydrates through the glyoxylate cycle [46]. It was also proposed that fatty acids become oxidized to carbon dioxide and water to generate energy for germination, leading to the synthesis of certain structural components in young seedlings [42].

The ash and fiber content in lima beans increased from 3.78 (RG) to 4.49% (G72 h) and 6.55 (RG) to 7.94% (G72 h), respectively. The ash and fiber content increased by 18.78 and 21.22%, respectively, after germination. Similarly, the value of the ash and fiber contents in adzuki beans increased by 18.54 and 29.01%, respectively. The increase in ash content may be apparent due to the loss of starch in germination. Ejigui et al. [46] reported an 11% increase in ash content during germination in red kidney beans. Devi et al. [47] reported a 30.46% increase in fiber content during germination in cowpea seeds. The crude fiber consisting of hemicelluloses, lignin, and cellulose enhanced significantly during sprouting as the plant cells synthesized various cellular compounds [48].

The carbohydrate content and calorific value in lima beans decreased from 62.03 (RG) to 56.89% (G72 h) and 350.96 (RG) to 335.95% (G72 h), respectively. The carbohydrate content and calorific value in lima beans decreased by 8.28 and 4.27%, respectively, after soaking and germination. Similarly, the carbohydrate and calorific values in adzuki beans decreased by 20.93 and 11.91%, respectively, after soaking and germination treatments. Similar results were found by Kavitha and Parimalavalli [49], who reported a 20.06 and 9.42% decrease in carbohydrate content and calorific value, respectively, in germinated maize flour. Vidal-Valverde et al. [50] reported that during sprouting, carbohydrate content was used as a source of energy for embryonic growth.

The in vitro protein digestibility (IVPD) in lima and adzuki beans increased by 14.75 and 10.98%, respectively. The results were comparable with the findings of Alonso et al. [32], who observed a 14.53% increase in in vitro protein digestibility in *Phaseolus vulgaris*. Another study by Sharma et al. [51] reported an 11.62% increase in the IVPD in pigeon peas. The increase in IVPD in all process treatments might be due to a reduction in anti-nutrients and disintegration in the structure of some native proteins, including the enzyme inhibitors and lectins. It may be the result of increased phytase activity causing the breakdown of phytic acid in the seeds and augmented α-galactosidase activity to reduce oligosaccharides in seeds. Also, the breakdown of proteins into amino acids resulted in increased digestibility of proteins [52].

3.5.2. Anti-Nutritional Components

The phytic and tannin content in lima beans decreased from 20.48 (RG) to 12.68 (G72) and 10.70 (RG) to 4.35 mg/100 g (G72), respectively. The phytic and tannin contents in adzuki beans decreased from 7.12 (RG) to 3.45 (G72) and 6.36 (RG) to 2.47(G72), respectively. The reduction in phytic and tannin contents in lima beans was 38.08 and 59.34%, respectively, and the same in adzuki beans was reported as 51.54 and 61.16%, respectively, after germination for 72 h. Yasmin et al. [53] found reductions of 42.62 and 68.85% in phytic and tannin content in germinated kidney beans, respectively. Patterson et al. [54] also found a 43% decline in the tannin content in chickpeas after 24 h of germination treatment, and Olika et al. [55] found a 57.35% decrease in phytic acid content in germinated chickpeas. This decline in phytic acid contents can be attributed to an increase in phytase activity during sprouting [56]. The decrease in tannin content after sprouting occurs due to the formation of a hydrophobic association of tannins with seed proteins and enzymes and the leaching of tannins into the water due to a concentration gradient [57].

The trypsin inhibitor activity (TIA) in lima and adzuki beans decreased to 84.22 and 71.01%, respectively, after germination treatment. Jayalaxmi et al. [43] observed a 77.57% decrease in TIA in germinated lima beans. Another report by Patterson et al. [54] found a 76% reduction in TIA in white kidney beans after 5 days of germination. The reduction in TIA after the sprouting might be due to the increased proteolytic activity of enzymes, which become activated during germination, and TIA decreases with an increase in trypsin activity [58].

3.5.3. Antioxidant Activity, Flavonoids, Phenols, Reducing Capacity, and Metal Chelating Activity

There was a significant increase in polyphenols during germination, which contributed significantly to increasing antioxidant activities. The antioxidant activity in lima beans increased from 35.48 (RG) to 47.36% (G72) and 9.74 (RG) to 16.67% (G72) in adzuki beans. The total increase in antioxidant activity in lima and adzuki beans was 33.48 and 71.14%, respectively. Yu-Wei and Wang [37] found an increase of 62.29% in antioxidant activity in adzuki beans. The antioxidant components increased due to the amount of phenolic content that increased in sprouted legumes due to the presence of various hydroxyl groups that behaved like free radical scavengers and resulted in an increase in the antioxidant activity of germinated grains; some bioactive compounds were also found to become increased, and these bioactive compounds acted as an additional antioxidant compound for increasing the antioxidant activity in sprouted beans [59].

The flavonoid content in lima and adzuki beans increased from 20.16 (RG) to 33.63 mg QE/100 g (G72) and 25.38 (RG) to 34.47 mgQE/100 g (G72), respectively. There were 66.81 and 35.81% increases in flavonoid content in lima and adzuki beans, respectively, during 72 h of germination. The results were comparable with the findings of Sharma et al. [46], who reported a 60.14% increase in flavonoid content in germinated pigeon peas. Kaur et al. [60] also found a 21.85% increase in flavonoid content in rice beans during germination.

The reducing capacity in lima and adzuki beans increased from 30.55 (RG) to 45.68% (G72) and 28.77 (RG) to 39.25% (G72), respectively. There were 49.52 and 36.42% increases in the reducing capacity of lima and adzuki beans, respectively, during 72 h of germination. Similar results were found by James et al. [61], who reported a 57.14% increase in reducing capacity in germinated red beans. Khang et al. [39] also found a 42.51% rise in reducing capacity after 5 days of germination. Similarly, the metal chelating activity in lima and adzuki beans increased by 56.06 and 68.23%, respectively. Sharma et al. [51] reported a 62.97% increase in metal chelating activity in germinated pigeon peas. Phytic acid, as phytate, forms strong complexes with many metal ions, thus competing in complexation with EDTA. An increase in the metal chelating activity can be attributed to the increase in the availability of metal ions due to a reduction in phytic acid mass fraction [62].

The phenolic content in lima and adzuki beans increased from 20.48 (RG) to 12.68 mg/g (G72) and 16.16 (RG) to 22.39% (G72), respectively. There were 7.13 and 38.55% increases in the phenolic content of lima and adzuki beans, respectively, during 72 h of germination. Sharma et al. [45] found a 4.7% increase in phenolic content in germinated pigeon peas. Khang et al. [39] also observed a 30.13% increase in phenolic content in adzuki beans. This could be due to the fact that the germinated seeds improved their defensive response by biosynthesis of phenolic compounds, resulting in their survival during germination [63]. Protease also activates and hydrolyzes the protein in the seed during germination. Also, some protein-bound phenolic components were solubilized.

3.5.4. Mineral Contents

The zinc content in lima and adzuki beans increased from 2.45 (RG) to 2.76 ppm (G72) and 4.49 (RG) to 6.74 ppm (G72), respectively, representing 11.83 and 50.11% increase in Zn content of lima and adzuki beans during 72 h of germination, respectively. Similar results were found by Farinde et al. [36] and Kajla et al. [64], who reported a 13.77% increase in zinc content in germinated lima beans and a 50.53% increase in zinc content in germinated flaxseed.

The copper content in lima and adzuki beans increased from 4.17 (RG) to 5.27 ppm (G72) and 1.22 (RG) to 1.51 ppm (G72), respectively, with an increase of 26.37 and 23.77% in Cu content of lima and adzuki beans during 72 h of germination. Similarly, the manganese content increased from 2.36 (RG) to 3.55 ppm (G72) and 1.57 (RG) to 2.22 ppm (G72), respectively. There was an increase of 50.42 and 41.40% in Mn in lima and adzuki beans during 72 h of germination, respectively.

The iron content of the lima and adzuki beans enhanced from 4.04 (RG) to 4.96 ppm (G72) and 5.16 (RG) to 7.32 ppm (G72), respectively, during 72 h of germination. This represents a 23.01 and 41.86% increase in Fe content, respectively. Kajla et al. [64] reported a 15.44% increase in copper content, 37.27 in manganese, and a 37.22% increase in zinc content in germinated flaxseeds. The increase in mineral content during germination and soaking may be due to partial elimination or destruction or both of the anti-nutritional factors, thereby releasing the minerals from their organically bound complexes in the dry seeds [65].

3.5.5. Fourier-Transform Infrared (FTIR) Spectroscopic Analysis

The FTIR spectra of soaking and germination treatments of adzuki beans are given in Figure 3 and Table 5.

Table 5. Band assignment of raw, soaked, and germinated grains of adzuki and lima beans.

Wave Range (cm^{-1})	Functional Group	Compound Class	Adzuki Beans			Lima Beans		
			Raw Grains	Soaking	Germination	Raw Grains	Soaking	Germination
3400–3300	N-H stretching	Aliphatic primary amine	3272	3290	3287	3260	3266	3262
3300–2500	O-H stretching	Carboxylic acid	2905	2914	2913	2911	2922	2922
1550–1500	N-O stretching	Nitro compound	1631	1632	1635	1625	1635	1631
980–960	C=C bending	Alkene	992	996	994	997	994	994
600–500	C-I stretching	Halo compound	519	515	518	482	522	515

The appearance of OD band is shown in the region of 3400–3300, 3300–2500, 1550–1500, 980–960, and 600–500 cm^{-1} in the three FTIR spectra, which shows that these have N-H, O-H, C=O, N-O, C=C, and C-I functional groups, respectively. The N–H stretching with wave no. of 3262, 3266, and 3260 cm^{-1} was reported in raw grain, soaked, and germinated grains, respectively.

Additionally, a peak in absorbance was formed in the wave range of 2500–3300, 1550–1500, 980–960, and 600–500 cm^{-1} as a result of the characteristic O-H stretching, N-O stretching, C=C bending, and C-I stretching, respectively. From the present study, it was observed that soaked grains had higher N-H, O-H, N-O, and C-I stretching compared to C=C bending, which was observed highest in germinated seeds of lima bean grains. It was also observed that raw, germinated, and soaked seeds possessed similar chemical structures. In the carboxylic region, absorption leads to the stretching of the O=H bond, whereas in the aliphatic primary amine region, absorption leads primarily to the stretching of the N–H bond.

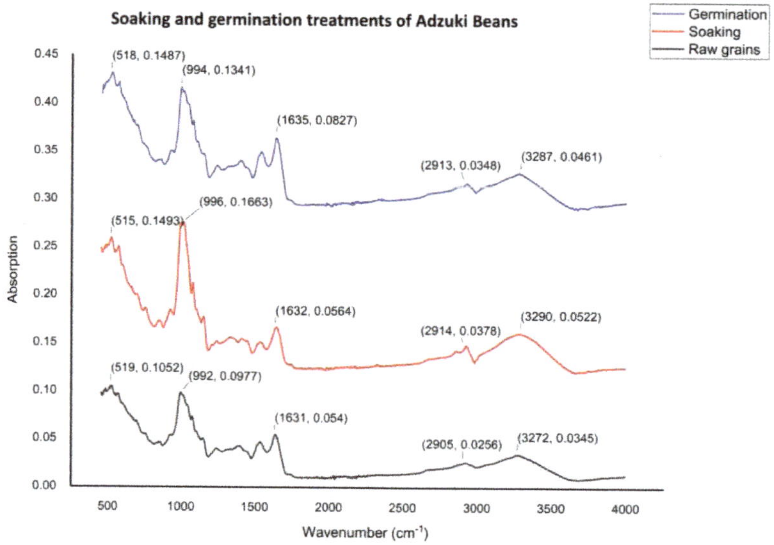

(**a**)

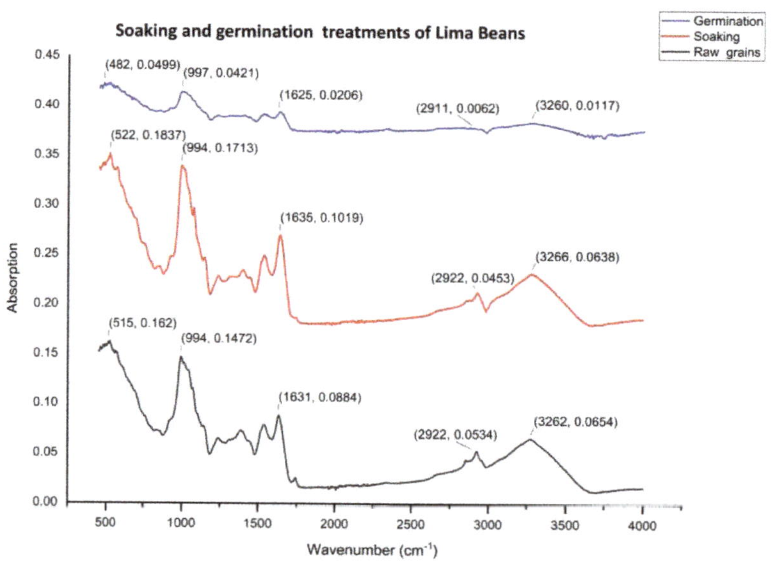

(**b**)

Figure 3. FTIR spectrum of adzuki (**a**) and lima beans (**b**) treated with soaking and germination treatments.

4. Conclusions and Future Aspects

The current study revealed the impact of processing treatments on the nutritional, anti-nutritional, and bioactive potential of adzuki and lima beans and concluded that there was a significant increase in proteins, minerals, phenolic compounds, in vitro protein digestibility, metal chelating activity, and reducing capacity. Processing treatments like soaking and germination do not require complicated types of equipment, rendering them simple techniques to improve the nutritional and sensory characteristics of germinated grains. Soaking could be one of the processes to remove soluble anti-nutritional factors, which can be eliminated with the discarded soaking solution. During germination, the anti-nutritional factors such as tannins, trypsin inhibitors, and phytic acids decreased with an increase in germination time. Leaching of tannins during soaking and their further degradation during germination and increase in the activity of enzyme phytase resulted in the reduction in anti-nutritional components, thereby increasing mineral availability.

The flour obtained from the germinated beans can be incorporated into various functional bakery products, breakfast bars, snacks, and prebiotic non-dairy beverages, which can be used to eradicate protein deficiency among different sections of society. Due to their high nutritional and bioactive components, these have great potential for improving food and nutritional security and a sustainable and inexpensive meat alternative, especially for those from lower socioeconomic backgrounds and the world's predominately vegetarian and vegan communities. This can also encourage the scientific community, industry, and government to invest in research and development to increase germinated pulse-based foods. These processing methods can also be used to improve the nutritional and bioactive potential of other pulses while lowering the levels of anti-nutrients. This study will help promote soaking and germination processing as effective treatments for enhancing nutritional quality and reducing anti-nutrients, as well as the increased utilization of these underutilized pulses.

Author Contributions: Conceptualization, K.K.; methodology, Q.U.E.H.R.; software, M.A.S., N.A. and P.S.; validation, Q.U.E.H.R., K.K., N.A. and A.N.Y.; data curation, N.A.; writing—original draft preparation, Q.U.E.H.R. and K.K.; review and editing, R.P.F.G.; visualization, K.K., I.S. and M.A.S.; supervision, K.K.; project administration, K.K.; funding acquisition, R.P.F.G. All authors have read and agreed to the published version of the manuscript.

Funding: We received funding for the APC from the FCT—Foundation for Science and Technology (Portugal) through project Ref. UIDB/00681/2020. Furthermore, we would like to thank the CERNAS Research Centre (project DOI: 10.54499/UIDP/00681/2020) and the Polytechnic University of Viseu for their support. The support for this study from the Ministry of Food Processing Industries (MoFPI) Govt. of India for the creation of infrastructural facility (F. No. 5–11/2010-HRD) is gratefully acknowledged by the authors.

Institutional Review Board Statement: Not applicable.

Informed Consent Statement: Not applicable.

Data Availability Statement: The original contributions presented in the study are included in the article, further inquiries can be directed to the corresponding authors.

Conflicts of Interest: The authors declare that this article's content has no conflicts of interest.

References

1. Bhat, R.; Karim, A. Exploring the nutritional potential of wild and underutilized legumes. *Compr. Rev. Food Sci. Food Saf.* **2009**, *8*, 305–331. [CrossRef]
2. Ogechukwu, C.O.; Ikechukwu, I.O. Effect of heat processing treatments on the chemical composition and functional properties of lima bean (*Phaseolus lunatus*) flour. *Am. J. Food Sci. Nutr.* **2017**, *1*, 14–24. [CrossRef]
3. Yellavila, S.; Agbenorhevi, J.K.; Asibuo, J.; Sampson, G. Proximate composition, minerals content and functional properties of five lima bean accessions. *J. Food Secur.* **2015**, *3*, 69–74.
4. Agarwal, S.; Chauhan, E.S. Adzuki beans-physical and nutritional characteristics of beans and its health benefits. *Int. J. Health Sci. Res.* **2019**, *9*, 304–310.
5. Liu, R.; Zheng, Y.; Cai, Z.; Xu, B. Saponins and flavonoids from adzuki bean (*Vigna angularis* L.) ameliorate high-fat diet-induced obesity in ICR mice. *Front. Pharmacol.* **2017**, *8*, 687. [PubMed]
6. Embaby, H.E. Effect of heat treatments on certain antinutrients and in vitro protein digestibility of peanut and sesame seeds. *Food Sci. Technol. Res.* **2010**, *17*, 31–38. [CrossRef]
7. Singh, B.; Singh, J.P.; Shevkani, K.; Singh, N.; Kaur, A. Bioactive constituents in pulses and their health benefits. *J. Food Sci. Technol.* **2017**, *54*, 858–870. [CrossRef] [PubMed]

8. Kajihausa, O.; Fasasi, R.; Atolagbe, Y. Effect of different soaking time and boiling on the proximate composition and functional properties of sprouted sesame seed flour. *Niger. Food J.* **2014**, *32*, 8–15. [CrossRef]
9. Rizvi, Q.U.E.H.; Kumar, K.; Ahmed, N.; Yadav, A.N.; Chauhan, D.; Thakur, P.; Jan, S.; Sheikh, I. Influence of soaking and germination treatments on the nutritional, anti-nutritional, and bioactive composition of pigeon pea (*Cajanus cajan* L.). *J. Appl. Biol. Biotech.* **2022**, *10*, 127–134. [CrossRef]
10. Sangronis, E.; Machado, C. Influence of germination on the nutritional quality of Phaseolus vulgaris and Cajanus cajan. *LWT-Food Sci. Technol.* **2007**, *40*, 116–120. [CrossRef]
11. Chauhan, D.; Kumar, K.; Ahmed, N.; Thakur, P.; Rizvi, Q.U.E.H.; Jan, S.; Yadav, A.N. Impact of soaking, germination, fermentation, and roasting treatments on nutritional, anti-nutritional, and bioactive composition of black soybean (*Glycine max* L.). *J. Appl. Biol. Biotech.* **2022**, *10*, 186–192. [CrossRef]
12. Egli, I.; Davidsson, L.; Juillerat, M.; Barclay, D.; Hurrell, R. The influence of soaking and germination on the phytase activity and phytic acid content of grains and seeds potentially useful for complementary feedin. *J. Food Sci.* **2002**, *67*, 3484–3488. [CrossRef]
13. American Association of Cereal Chemists; Approved Methods Committee. *Approved Methods of the American Association of Cereal Chemists*; American Association of Cereal Chemists: St. Paul, MN, USA, 2000.
14. Huang, S.; Shiau, C.; Liu, T.; Chu, C.; Hwang, D.F. Effects of rice bran on sensory and physico-chemical properties of emulsified pork meatballs. *Meat Sci.* **2005**, *70*, 613–619. [CrossRef]
15. Jones, D.; Chinnaswamy, R.; Tan, Y.; Hanna, M. Physiochemical properties of ready-to-eat breakfast cereals. *Cereal Foods World* **2000**, *45*, 164–168.
16. Sosulski, F. The centrifuge method for determining flour absorption in hard red spring wheats. *Cereal Chem.* **1962**, *39*, 344–350.
17. Kaur, M.; Sandhu, K.S.; Arora, A.; Sharma, A. Gluten free biscuits prepared from buckwheat flour by incorporation of various gums: Physicochemical and sensory properties. *LWT-Food Sci. Technol.* **2015**, *62*, 628–632. [CrossRef]
18. Stojceska, V.; Ainsworth, P.; Plunkett, A.; İbanoğlu, E.; İbanoğlu, Ş. Cauliflower by-products as a new source of dietary fibre, antioxidants and proteins in cereal based ready-to-eat expanded snacks. *J. Food Eng.* **2008**, *87*, 554–563. [CrossRef]
19. AOAC. *Official Methods of Analysis*; AOAC: Washington, DC, USA, 1990.
20. Ranganna, S. *Handbook of Analysis and Quality Control for Fruit and Vegetable Products*; Tata McGraw-Hill Education: Noida, India, 1986.
21. WHO. *Energy and Protein Requirements*; WHO Technical Report Series no 522 and FAO Nutrition Meeting Report Series no 52 Geneva; World Health Organization: Geneva, Switzerland, 1973.
22. Shastry, M.; John, E. Biochemical changes and in-vitro protein digestibility of the endosperm of germinating dolichos lablab. *J. Sci. Food Agric.* **1991**, *55*, 529–538. [CrossRef]
23. Lahlou, F.A.; Hmimid, F.; Loutfi, M.; Bourhim, N. Antioxidant activity and quantification of phenolic compounds of Euphorbia echinus. *Int. J. Pharm. Sci.* **2014**, *6*, 357–360.
24. Bouaziz, M.; Fki, I.; Jemai, H.; Ayadi, M.; Sayadi, S. Effect of storage on refined and husk olive oils composition: Stabilization by addition of natural antioxidants from Chemlali olive leaves. *Food Chem.* **2008**, *108*, 253–262. [CrossRef]
25. Sharma, P.; Gujral, H.S. Effect of sand roasting and microwave cooking on antioxidant activity of barley. *Food Res. Int.* **2011**, *44*, 235–240. [CrossRef]
26. Saxena, V.; Mishra, G.; Saxena, A.; Vishwakarma, K. A comparative study on quantitative estimation of tannins in *Terminalia chebula, Terminalia belerica, Terminalia arjuna* and *Saraca indica* using spectrophotometer. *Asian J. Pharm. Clin. Res.* **2013**, *6*, 148–149.
27. Kakade, M.; Rackis, J.; McGhee, J.; Puski, G. Determination of trypsin inhibitor activity of soy products: A collaborative analysis of an improved procedure. *Cereal Chem.* **1974**, *51*, 376–382.
28. Gao, Y.; Shang, C.; Maroof, M.S.; Biyashev, R.; Grabau, E.; Kwanyuen, P.; Burton, J.; Buss, G. A modified colorimetric method for phytic acid analysis in soybean. *Crop Sci.* **2007**, *47*, 1797–1803. [CrossRef]
29. Ainsworth, E.A.; Gillespie, K.M. Estimation of total phenolic content and other oxidation substrates in plant tissues using Folin–Ciocalteu reagent. *Nat. Protoc.* **2007**, *2*, 875–877. [CrossRef]
30. Farinde, E.O.; Olanipekun, O.T.; Olasupo, R.B. Nutritional composition and antinutrients content of raw and processed lima bean (*Phaseolus lunatus*). *Ann. Food Sci. Technol* **2018**, *19*, 250–264.
31. Daba, H.G.; Delele, M.A.; Fanta, S.W.; Habtu, N.G.; Abera, M.K.; Worku, A.F. Effects of Initial Moisture Content and Storage Duration on Physical and Chemical Characteristics of Stored Maize (*Zea mays* L.) Grain. In *Advancement of Science and Technology in Sustainable Manufacturing and Process Engineering*; Springer: Cham, Switzerland, 2023; pp. 109–117, ISBN 978-3-031-41172-4.
32. Purwanti, E.; Fauzi, A. The morphological characteristics of *Phaseolus lunatus* L. in different areas of East Java, Indonesia. In *IOP Conference Series: Earth and Environmental Science, Proceedings of the International Conference on Life Sciences and Technology, Malang, Indonesia, 4 September 2018*; IOP Publishing: Bristol, UK, 2019; Volume 276, p. 012017.
33. Wu, W.; Zhou, L.; Chen, J.; Qiu, Z.; He, Y. Gain, TKW: A measurement system of thousand kernel weight based on the android platform. *Agronomy* **2018**, *8*, 178. [CrossRef]
34. Yadav, U.; Singh, N.; Kaur, A.; Thakur, S. Physico-chemical, hydration, cooking, textural and pasting properties of different adzuki bean (*Vigna angularis*) accessions. *J. Food Sci. Technol.* **2018**, *55*, 802–810. [CrossRef]
35. Siddiq, M.; Ravi, R.; Harte, J.; Dolan, K. Physical and functional characteristics of selected dry bean (*Phaseolus vulgaris* L.) flours. *LWT-Food Sci. Technol.* **2010**, *43*, 232–237. [CrossRef]

36. Woo, K.S.; Song, S.B.; Ko, J.Y.; Kim, Y.B.; Kim, W.H.; Jeong, H.S. Antioxidant properties of adzuki beans, and quality characteristics of sediment according to cultivated methods. *Korean J. Food Nutri.* **2016**, *29*, 134–143. [CrossRef]
37. Yu-Wei, L.; Wang, Q. Effect of processing on phenolic content and antioxidant activity of four commonly consumed pulses in China. *J. Hortic.* **2015**, *2*, 130–132.
38. Alonso, R.; Aguirre, A.; Marzo, F. Effects of extrusion and traditional processing methods on antinutrients and in vitro digestibility of protein and starch in faba and kidney beans. *Food Chem.* **2000**, *68*, 159–165. [CrossRef]
39. Khang, D.T.; Dung, T.N.; Elzaawely, A.A.; Xuan, T.D. Phenolic profiles and antioxidant activity of germinated legumes. *Foods* **2016**, *5*, 27. [CrossRef] [PubMed]
40. Sombié, P.A.E.D.; Compaoré, M.; Coulibaly, A.Y.; Ouédraogo, J.T.; Tignégré, J.-B.D.L.S.; Kiendrébéogo, M. Antioxidant and phytochemical studies of 31 cowpeas (*Vigna unguiculata* (L. Walp.)) genotypes from Burkina Faso. *Foods* **2018**, *7*, 143. [CrossRef]
41. Adeparusi, E. Effect of processing on the nutrients and anti-nutrients of lima bean (*Phaseolus lunatus* L.) flour. *Food/Nahrung* **2001**, *45*, 94–96. [CrossRef] [PubMed]
42. Granito, M.; Brito, Y.; Torres, A. Chemical composition, antioxidant capacity and functionality of raw and processed Phaseolus lunatus. *J. Sci. Food Agric.* **2007**, *87*, 2801–2809. [CrossRef]
43. Jayalaxmi, B.; Vijayalakshmi, D.; Usha, R.; Revanna, M.; Chandru, R.; Gowda, P. Effect of different processing methods on proximate, mineral and antinutrient content of lima bean (*Phaseolus lunatus*) seeds. *Legume Res.-Int. J.* **2016**, *39*, 543–549. [CrossRef]
44. Mubarak, A. Nutritional composition and antinutritional factors of mung bean seeds (*Phaseolus aureus*) as affected by some home traditional processes. *Food Chem.* **2005**, *89*, 489–495. [CrossRef]
45. Nonogaki, H.; Bassel, G.W.; Bewley, J.D. Germination—Still a mystery. *Plant Sci.* **2010**, *179*, 574–581. [CrossRef]
46. Ejigui, J.; Savoie, L.; Marin, J.; Desrosiers, T. Influence of traditional processing methods on the nutritional composition and antinutritional factors of red peanuts (*Arachis hypogea*) and small red kidney beans (*Phaseolus vulgaris*). *J. Biol. Sci.* **2005**, *5*, 597–605.
47. Devi, C.B.; Kushwaha, A.; Kumar, A. Sprouting characteristics and associated changes in nutritional composition of cowpea (*Vigna unguiculata*). *J. Food Sci. Technol.* **2015**, *52*, 6821–6827. [CrossRef] [PubMed]
48. Laxmi, G.; Chaturvedi, N.; Richa, S. The impact of malting on nutritional composition of foxtail millet, wheat and chickpea. *J. Nutr. Food Sci.* **2015**, *5*, 5.
49. Kavitha, S.; Parimalavalli, R. Effect of processing methods on proximate composition of cereal and legume flours. *J. Hum. Nutr. Food Sci.* **2014**, *2*, 1051.
50. Vidal-Valverde, C.; Frias, J.; Sierra, I.; Blazquez, I.; Lambein, F.; Kuo, Y.-H. New functional legume foods by germination: Effect on the nutritive value of beans, lentils and peas. *Eur. Food Res. Technol.* **2002**, *215*, 472–477. [CrossRef]
51. Sharma, S.; Singh, A.; Singh, B. Characterization of in vitro antioxidant activity, bioactive components, and nutrient digestibility in pigeon pea (*Cajanus cajan*) as influenced by germination time and temperature. *J. Food Biochem.* **2019**, *43*, e12706. [CrossRef]
52. Sinha, R.; Kawatra, A.; Sehgal, S. Saponin content and trypsin inhibitor activity of cowpea: Varietal differences and effects of processing and cooking methods. *J. Food Sci. Technol.* **2005**, *42*, 182–185.
53. Yasmin, A.; Zeb, A.; Khalil, A.W.; Paracha, G.M.; Khattak, A.B. Effect of processing on anti-nutritional factors of red kidney bean (*Phaseolus vulgaris*) grains. *Food Bioproc. Tech.* **2008**, *1*, 415–419. [CrossRef]
54. Patterson, C.A.; Curran, J.; Der, T. Effect of processing on antinutrient compounds in pulses. *Cereal Chem.* **2017**, *94*, 2–10. [CrossRef]
55. Olika, E.; Abera, S.; Fikre, A. Physicochemical properties and effect of processing methods on mineral composition and antinutritional factors of improved chickpea (*Cicer arietinum* L.) varieties grown in Ethiopia. *Int. J. Food Sci.* **2019**, *2019*, 9614570. [CrossRef]
56. Khattak, A.B.; Zeb, A.; Bibi, N.; Khalil, S.A.; Khattak, M.S. Influence of germination techniques on phytic acid and polyphenols content of chickpea (*Cicer arietinum* L.) sprouts. *Food Chem.* **2007**, *104*, 1074–1079. [CrossRef]
57. Shimelis, E.A.; Rakshit, S.K. Effect of processing on antinutrients and in vitro protein digestibility of kidney bean (*Phaseolus vulgaris* L.) varieties grown in East Africa. *Food Chem.* **2007**, *103*, 161–172.
58. Chauhan, O.; Chauhan, G. Development of anti-nutrients free soy beverage using germinated soybean. *J. Food Sci. Technol.* **2007**, *44*, 62–65.
59. Uchegbu, N.N.; Ishiwu, C.N. Germinated Pigeon Pea (*Cajanus cajan*): A novel diet for lowering oxidative stress and hyperglycemia. *Food Sci. Nutr.* **2016**, *4*, 772–777. [CrossRef]
60. Kaur, D.; Dhawan, K.; Rasane, P.; Singh, J.; Kaur, S.; Gurumayum, S.; Singhal, S.; Mehta, C.; Kumar, V. Effect of different pre-treatments on antinutrients and antioxidants of rice bean (*Vigna umbellata*). *Acta Univ. Cinbinesis Ser. E Food Technol.* **2020**, *24*, 25–38. [CrossRef]
61. James, S.; Nwabueze, T.U.; Ndife, J.; Onwuka, G.I.; Usman, M.A.A. Influence of fermentation and germination on some bioactive components of selected lesser legumes indigenous to Nigeria. *J. Agric. Food Res.* **2020**, *2*, 100086. [CrossRef]
62. Sharma, S.; Sahni, P. Dynamics of germination behaviour, protein secondary structure, technofunctional properties, antinutrients, antioxidant capacity and mineral elements in germinated dhaincha. *Food Technol. Biotechnol.* **2021**, *59*, 238–250. [CrossRef] [PubMed]
63. Randhir, R.; Lin, Y.T.; Shetty, K. Stimulation of phenolics, antioxidant and antimicrobial activities in dark germinated mung bean sprouts in response to peptide and phytochemical elicitors. *Process Biochem.* **2004**, *39*, 637–646. [CrossRef]

64. Kajla, P.; Sharma, A.; Sood, D.R. Effect of germination on proximate principles, minerals and anti nutrients of flaxseeds. *Asian J Dairy Food Res.* **2017**, *36*, 52–57. [CrossRef]
65. Obizoba, I.C. Effect of sprouting on the nitrogenous constituents and mineral composition of pigeon pea (*Cajanus cajan*) seeds *Plant Foods Hum. Nutr.* **1991**, *41*, 21–26. [CrossRef]

foods

MDPI

Article

Food Security in Israel: Challenges and Policies

Ayal Kimhi [1,2]

[1] Shoresh Institution for Socioeconomic Research, Kochav Yair 44864, Israel; ayal@shoresh.institute or ayal.kimhi@mail.huji.ac.il
[2] Department of Environmental Economics and Management, Institute of Environmental Sciences, The Robert H. Smith Faculty of Agriculture, Food and Environment, The Hebrew University of Jerusalem, Rehovot 76100, Israel

Abstract: This article analyzes Israel's food security in comparison to other developed countries, using multiple indicators divided into four sections: food availability, food affordability, food quality and safety, and natural resources and resilience. Overall, the state of food security in Israel is better than in most countries, but the threats to food security arising from the triple risk of climate change, international conflicts, and disruptions in global supply chains, require better preparation for the future. Israel's population growth and the slowdown in the growth rate of its agricultural production, as well as the short-term political desire to reduce prices, are leading the country to increasingly rely on food imports. Such imports expose Israel to even greater global risks, and require the formulation of a risk-management strategy that will balance local production and imports. The global triple risk to food security is currently exacerbated for Israel by the risk of shortage of labor due to the security situation, making this risk-management strategy even more necessary. This calls for the establishment of a governmental authority to oversee the formulation of a long-term food-security strategy, to break it down into feasible objectives and policy measures, and to supervise their implementation. Most importantly, in order to maintain and perhaps even enhance the productive capacity of the agricultural sector, the government must reinstall trust between farmers and the state by establishing a stable long-term policy environment.

Keywords: food security; Israel; triple risk; risk-management strategy

1. Introduction

The internationally accepted definition of food security is that all people, at all times, have physical, social, and economic access to sufficient, safe, and nutritious food that meets their food preferences and dietary needs for an active and healthy life [1]. Measuring the level of food security, at a global level, at a country level, at a household level, or at an individual level, is a non-negligible challenge, but all governments, and many NGOs, see themselves as having an obligation to ensure the maximum level of food security for the citizens of their country. Food security is an important component of national security, and its promotion requires a combination of policy measures in the fields of welfare, health, agriculture, environmental quality, and international trade. Modern food systems have increasingly succeeded in improving the access of individuals and households to food, in terms of both quantity and price. But, they have been less successful in improving the nutritional and health value of the food basket, in protecting the environment [2], and in allowing all population groups to benefit equally from the improvements [3].

In the first years following Israel's independence, there was not enough food in the country to feed the rapidly growing population resulting from the massive waves of immigration. Consequently, the government had to implement food rationing. However, massive investments in agriculture bore fruit. In Israel's first four decades, the quantity of agricultural output grew faster than the country's population (Figure 1). However, as Figure 1 shows, the growth rate of agricultural output is slowing down gradually, as in many developed countries [4], and is now lower than the rate of population growth. The slowdown in the growth of agricultural output stems from multiple causes. One is the loss of the most fertile farmland to urbanization and the move to more marginal land. Another cause is the shortage of water. Israel now desalinates water, but the cost of desalinated water is way too high for many crops. Farms rely more heavily on treated water, which is cheaper, but it is not suitable for all crops and may lead to yield losses in some crops. Environmental and health regulation of the use of pesticides and herbicides has tightened over the years, adversely affecting both yields and profitability. In addition, government support of agriculture has been declining over the years, and government policy in general has become less favorable to the farm sector.

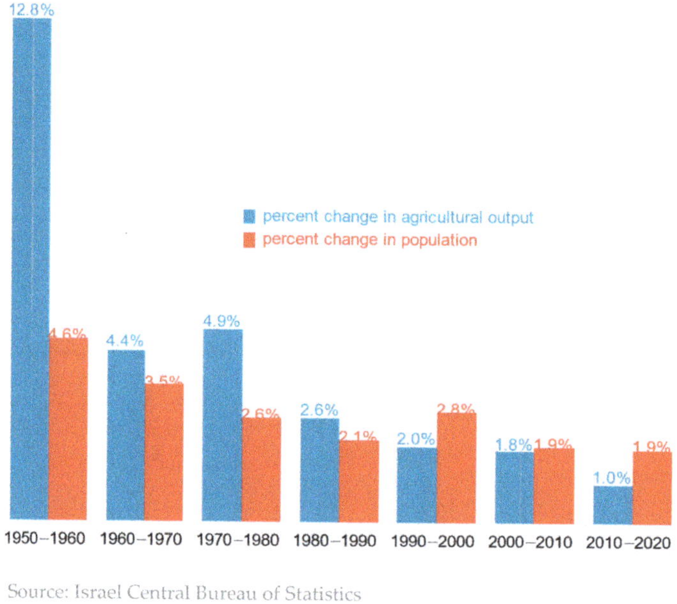

Figure 1. Annual increase in population and agricultural output.

Despite that, since the 1950s, Israel has never again experienced food shortages. During the 1970s, food prices in Israel moved closely with the OECD average (Figure 2). The high food prices during the 1980s were due to the hyperinflation and economic instability that Israel experienced at that time [5]. Food prices declined during the 1990s following the stabilization plan of 1985. The next time food prices entered the public discourse was in 2007–2008, when the prices of agricultural commodities in international markets soared and led to a significant increase in the price of food in Israel as well (Figure 2). The real increase in Israeli food prices was higher than the average real increase in the price of food in OECD countries. This contributed to the outbreak of social protests over the high cost-of-living in Israel in the summer of 2011.

The Commission for Economic and Social Change (the "Trachtenberg Commission") was established following the social protests. Its report listed the food sector among the sectors needing a price-reduction policy. At the same time, an "Inter-ministerial Commission for Examining the Level of Competitiveness and Prices in the Markets for Food and Consumer Products" ("Kedmi Committee") was established to study the characteristics of the food- and consumer-products markets, to locate market failures, if any, and to formulate recommendations for improving consumer welfare.

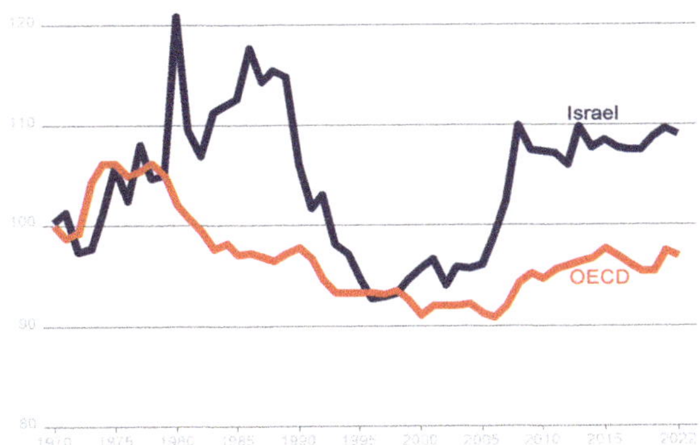

Source: OECD

Figure 2. Real-food-price index, Israel and OECD average.

The Kedmi committee report [6] placed most of the responsibility for the sharp increase in food prices that began in 2005 on the low competitiveness of the food sector. Competitiveness in the food sector declined significantly in 2005 with the purchase of the Clubmarket, the third-largest retail chain, by Shufersal, which resulted in an increase in the market share of the two largest chains (Shufersal and Mega). Market concentration allowed the large chains to raise food prices when producer prices rose due to the increase in the price of agricultural inputs, and to not lower food prices when producer prices fell (asymmetric price transmission) so that food prices remained high even after 2008, when the global recession led to a sharp drop in input prices and the exchange rate also fell, which should have made imported food cheaper. The committee's policy recommendations focused on reducing import tariffs on imported food products and actions in the consumer sector, as well as actions to reduce market concentration. The bulk of these recommendations were not implemented.

Discourse on food security followed the social protests focused on food prices. However, disruptions in global supply chains following the outbreak of the COVID pandemic in 2020 turned the spotlight on the danger of possible food shortages in Israel. For the first time in decades, a serious public discussion began on the limitations of globalization and the importance of self-production of basic consumer goods. This discussion intensified with the onset of Russia's war on Ukraine, as both countries are large exporters of both energy and food. The disruptions to these countries' exports caused global price increases of agricultural commodities such as wheat and sunflower oil, as well as various energy products. Since modern agriculture needs a considerable amount of energy for its production and transport, higher energy prices lead to food-price increases. Finally, the onset of Israel's war against the Hamas terrorist organization has led to offensive verbal responses from leaders of neighboring Islamic countries that are important sources of imported produce to Israel, causing much concern in Israel about the long-term reliability of imports from these countries.

In light of the emerging threats to food security in Israel, several key questions arise:

1. How does Israel's food security compare internationally?
2. To what extent can (or should) Israel rely on self-production of its food?
3. To what extent does the rise in food prices, and especially the prices of fresh agricultural produce, pose a threat to food security?
4. How successful are existing policy measures in dealing with these threats, and what is the potential of other policy measures?

This study addresses these questions, primarily within the context of a multi-indicator food-security index. The previous literature mostly dealt with a narrower list of indicators. Endeweld and Silber [7] examined the historical development of the supply of food at the macro level and also the nutritional insecurity at the micro level. Griver and Fischhendler [8] show that Israeli food-security policy changed focus over the years following global and local events. Tal [9] discusses the role of public investments in research, development, and extension of enhancing food security in Israel. None of these authors offered an overall inclusive discussion on food security in Israel, especially not the future threats and the role of contemporary policy changes. This paper also discusses the unique situation of Israeli agriculture in light of the current armed conflict, and highlights the need for a long-term risk-management strategy for food security.

2. Methodology

This study utilizes an index of food security developed by the Economist Group. The index is constructed as a weighted average of a long list of elements affecting food security at a national level [10]. This index, compiled with the help of a team of experts, is divided into four main sections: food availability, food affordability, food quality and safety, and natural resources and resilience. Each of these sections includes a variable number of measures (Table 1), and the overall food-security index is a weighted average of all measures.

Table 1. Israel's ranking in the various components of the food-security index, 2021.

Food-Security Indicators [1]	Weight	Israel's Rank
Availability	**32.40%**	**5**
Sufficiency of supply	**26.30%**	**2**
Agricultural R&D	**9.10%**	**32**
Public expenditure on agricultural R&D	50.00%	15
Access to agricultural technology, education, and resources	50.00%	32
Agricultural infrastructure	**14.10%**	**24**
Road infrastructure	35.70%	24–30
Air, port, and rail infrastructure	35.70%	22
Irrigation infrastructure	28.50%	2
Volatility of agricultural production	**15.20%**	**7**
Political and social barriers to access	**12.10%**	**28**
Armed conflict	29.40%	26–29
Political stability risk	23.50%	27
Corruption	23.50%	19–28
Gender inequality	23.50%	19
Food loss	**14.10%**	**9**
Food-security and -access policy commitments	**9.10%**	**1–17**
Food-security strategy	50.00%	18–32
Food-security council	50.00%	1–3
Affordability	**32.40%**	**7**
Change in average food cost	**29.80%**	**2**
Proportion of population under global poverty line [2]	**27.00%**	**20**
Inequality-adjusted income index [3]	**29.80%**	**27**
Agricultural-import tariffs	**13.60%**	**24**
Quality and safety	**17.60%**	**10**
Dietary diversity	**20.30%**	**24**
Nutritional standards	**13.60%**	**13–21**
Micronutrient availability	**25.40%**	**2**
Protein quality	**23.70%**	**1**
Food safety	**16.90%**	**1–8**

Table 1. *Cont.*

Food-Security Indicators [1]	Weight	Israel's Rank
Natural resources and resilience	**17.60%**	**31**
Exposure	**21.10%**	**9**
Temperature rise	27.30%	10
Drought	25.00%	23–32
Flooding	22.70%	3
Sea level rise	25.00%	14
Water	**14.00%**	**15–25**
Agricultural water risk—quantity	80.00%	15–32
Agricultural water risk—quality	20.00%	1–19
Land	**14.00%**	**7**
Land degradation	60.00%	6
Grassland	20.00%	1
Forest change	20.00%	29
Oceans, rivers, and lakes	**12.30%**	**5**
Eutrophication	50.00%	4–32
Marine biodiversity	50.00%	3
Sensitivity	**10.50%**	**32**
Food-import dependency	60.00%	32
Dependence on natural capital	40.00%	13
Political commitment to adaptation	**21.10%**	**31**
Demographic stress	**7.00%**	**32**
Projected population growth	75.00%	32
Urban absorbment capacity	25.00%	31

[1] Primary components in bold; secondary components in black; and secondary subcomponents in gray. The sum of the weights of the components in each group or subgroup of components adds up to 100%. Source: [10]. [2] The proportion of the population whose daily income is less than $3.20 per day (at 2011 exchange rates adjusted for purchasing-power parity. [3] GNI per capita at 2011 PPP adjusted for level of inequality [11].

3. Food Security in Israel Compared to Other Countries

Figure 3 shows the food-security index of all countries for which the data are available as a function of GDP per capita. Not surprisingly, food security increases as GDP per capita increases. Specifically, the food-security index increases faster at lower levels of GDP per capita, i.e., among the poorest countries. The growth in the index declines as countries become wealthier. The solid line in the figure indicates a polynomial fit, and shows that Israel's food security is slightly better than what its GDP per capita predicts. Israel's food security is ranked in 12th place among the 32 OECD countries that participated in the ranking (the country ranked first enjoys the highest food security). However, a closer look at the components of the index provides a slightly less-optimistic picture about the future, as it pertains to food security in Israel. In terms of food availability and affordability, Israel ranks fifth and seventh, respectively. In terms of food quality and safety, Israel ranks tenth. These rankings represent the state of food security in Israel today.

However, in the field of natural resources and resilience, Israel is ranked second to last. This implies that if appropriate measures are not taken, food security in Israel is expected to deteriorate in the future. To gain a better understanding about what these rankings mean, this study delves into each of these components separately.

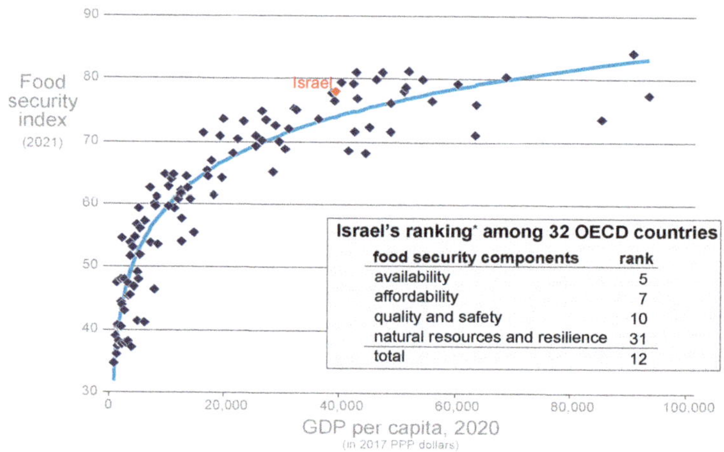

* The higher the rank, the higher food security index

Figure 3. Food-security index by GDP per capita among 110 countries, 2021. This figure is adapted with permission from Ayal Kimhi (2022). Copyright 2022; Copyright Ayal Kimhi. Source: Economist Impact (2021).

3.1. Food Availability

The food availability index includes components of sufficiency of supply, agricultural R&D, agricultural infrastructure, volatility of agricultural production, political and social barriers to access, food loss, and policy commitment (Table 1). Israel's high ranking in the field of food availability is primarily due to the fact that the country's current food supply is able to satisfy the energy requirements (in terms of calories) of the population and much more (Each person's energy requirement is the minimum number of calories they need to receive from food in order to fully function and to have an active immune system). However, in the area of agricultural R&D, which is intended to advance future food security, Israel receives much lower scores. On the one hand, public expenditure on agricultural R&D relative to the total agricultural product is about 42% of the total government investment as a share of GDP, which places Israel in the center of the distribution of the OECD countries (15th place among 32 countries). On the other hand, total-factor productivity (TFP) in Israel's agriculture sector increased by only 3.4% between 2012–2021, one of the lowest growth rates in the OECD. An increase in TFP reflects the ability to increase output without changing the quantities of inputs, while indirectly reflecting the contribution of agricultural R&D, which in Israel's case is relatively modest. (It should be noted that there is an academic debate about the methodology used to calculate TFP changes and the adaptation of this measure to the unique conditions of Israeli agriculture.)

Israel also ranks low in indicators of infrastructure quality. It is ranked at the bottom of the list of OECD countries in the areas of transport infrastructure, including roads, railways, and ports. On the other hand, it receives a high score in the field of irrigation infrastructure, since nearly half of its cultivated agricultural land is connected to an irrigation network. Israel ranks relatively high in the area of volatility in agricultural production. The reason for this may be the high percentage of land with an irrigation infrastructure, which makes crops less vulnerable to fluctuations in precipitation.

The rating of Israeli agriculture in terms of political and social barriers is quite low. This stems from the danger of armed conflict, political instability, corruption, and gender inequality. In the area of commitment to a food-security policy, Israel is ranked in the upper part of the distribution, apparently due to the existence of a national food-security council. However, it is ranked in the lower part of the distribution in the area of food-security strategy, probably because no such strategy exists in Israel [12].

As in other developed countries, about a third of the food produced in Israel does not reach the plate. Food loss occurs throughout the supply chain, but mainly at both ends: in consumers' homes and in agricultural farms [13]. However, food loss in Israel is not particularly high in comparison with other developed countries—with Israel ranking 9th in the OECD.

Ironically, food loss in consumers' homes might have been lower if food prices were higher. Consumers purchase food in larger quantities than necessary because there is uncertainty about the amount of food they will need. Increased food purchases can be viewed as "insurance" against a greater-than-expected demand for food. The cost of insurance is the cost of the food that is ultimately thrown away, and the higher the price of food, the less insurance consumers will be willing to "buy".

Another facet of this phenomenon is food that is thrown away by institutional consumers, such as banquet halls. Loss of food in agricultural farms is also to a large extent a result of the low prices of agricultural produce. When the price is too low, a farmer may make a decision (which is completely right for him) not to harvest the crop and thus save the cost of harvesting, sorting, and transportation. It follows that striving for zero food loss is impractical [14].

However, the decisions, which may be correct from the point of view of individual consumers and producers, are not necessarily the right ones at a national and global level. This is because the production of food, which is incompletely consumed, has environmental costs that are not taken into account by private agents (producers and consumers) [15]. Hence, countries have an interest in reducing food loss, often doing so through civil society organizations, which deal with saving agricultural produce and transferring it to those in need. However, such actions also have side effects. For example, distributing surplus food to the needy is likely to reduce their food purchases, which will lead to lower agricultural produce prices, and, as a result, larger quantities of crops may not be harvested.

3.2. Food Affordability

Food availability is measured using the change in the cost of food, the poverty rate, an inequality-adjusted income index, and import tariffs (Table 1). As shown in Figure 2, food prices in Israel rose during the 1980s (a period of three-digit inflation and a major debt crisis in agriculture, which drove many farms out of business), relative to food prices in other developed countries. This trend reversed in the 1990s, with food prices in Israel falling sharply during the decade. Food prices rose again in the following decade, and especially in the years 2005–2008, when world food prices rose substantially.

However, the increase in Israeli food prices was above and beyond the price increases in most OECD countries. An underlying reason for this may be the "Shufersal" chain's purchase of the "Clubmarket" retail chain, which was the third largest at the time. This purchase greatly increased the concentration on the food-retail sector. Since 2008, food prices in Israel have remained more or less stable. Overall, food prices in Israel increased since the end of the 1990s much more than in the OECD. It is possible that the strengthening of the Israeli currency (shekel) had a contribution to this, since the proportion of imports in the food basket of Israel is relatively high. Ben-David and Kimhi [16] found that compared to the total-consumption basket, food prices in Israel were about 3% lower than the OECD average in 2005, while in 2017 they were about 3% higher.

Prices of fruits and vegetables have increased in recent years compared to other food items. Figure 4 shows that the prices of fruits and vegetables moved closely with the price of the total food basket between 2000 and 2014. The steep rise in food prices during 2005–2008 reflects the trend in global food prices during that time. Since then, food prices experienced a modest rise over the years. However, starting in 2015, there was a steep increase in the prices of fruits and vegetables relative to the total food basket, reaching a maximum of 30% increase in 2020 compared to 2000. There are multiple reasons for this price rise, including a higher cost of production, tighter environmental regulations, larger produce variety, extensions of the cropping season, and, of course, climate-induced losses of output. The findings of Ben-David and Kimhi [16] indicate that this phenomenon is not unique to Israel. Fruits and vegetables became more expensive during this period in other developed countries as well. In fact, Ben-David and Kimhi [16] showed that prices of fruits and vegetables in Israel are lower than the average prices in the OECD countries when compared to the total consumption basket, both in 2005 and 2017. In addition, they showed that while the median wage in Israel enables the purchase of 15% fewer standard food baskets than the median wage in the OECD countries, it allows the purchase of 21% more fruit and vegetable baskets.

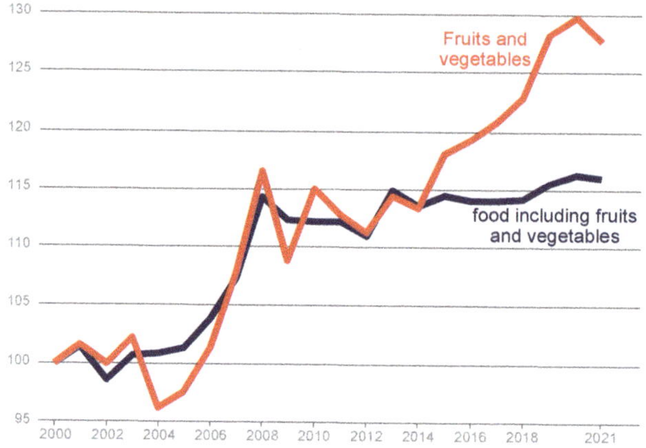

Source: Israel Central Bureau of Statistics

Figure 4. Real prices of food and of fruits and vegetables.

The fact that fruits and vegetables are more affordable in Israel than in other developed countries is perhaps relevant to the question of whether the removal of import barriers will succeed in reducing prices. Nonetheless, their price increase in recent years has been a concern for Israeli consumers. Fruits and vegetables are a significant component of the healthy food basket recommended by the Ministry of Health [17]. As such, their price increases hinder the recommended change in dietary habits. This conclusion is supported by Figure 5, which shows that the local per capita supply of fruits and vegetables in Israel has been declining since 2005. However, Israel's ranking in the food-security index places it in second place in the area of food prices. This is because food prices in Israel have increased at a moderate rate relative to most OECD countries since 2010. Specifically, food prices in Israel increased by 1.6% between 2010 and 2021, while the average price increase in the OECD countries was 2.4%. Conversely, import tariffs on agricultural products place Israel in 27th place in the OECD in this area, since they are seen as a factor that makes food more expensive.

Israel's poverty rates place it in 20th place among the OECD countries, and 27th in the area of income per capita adjusted for inequality (Table 1; Per capita income at 2011 prices according to purchasing power parity, adjusted for inequality according to the methodology in [10], which means that the greater the inequality, the smaller the adjusted income). Since Israel is one of the least equally developed countries [18], the concern for the public's ability to purchase a basket of healthy food is focused on the weaker populations. Azarieva and others [19] showed that in Israel, as in any other country, the share of food expenditures out of total household expenses rises as incomes fall. Specifically, 42% of the total expenses of the lowest income quintile (the fifth of all households with the lowest income) are spent on food (Figure 6).

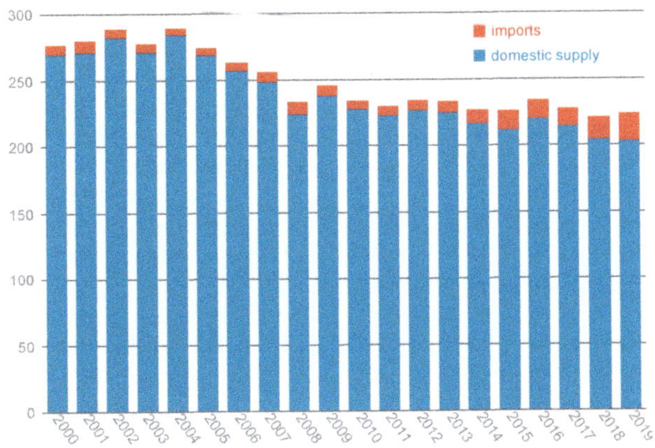

Source: Israeli Ministry of Agriculture and Rural Development

Figure 5. Supply of fruits and vegetables in Israel (annual kilograms per person).

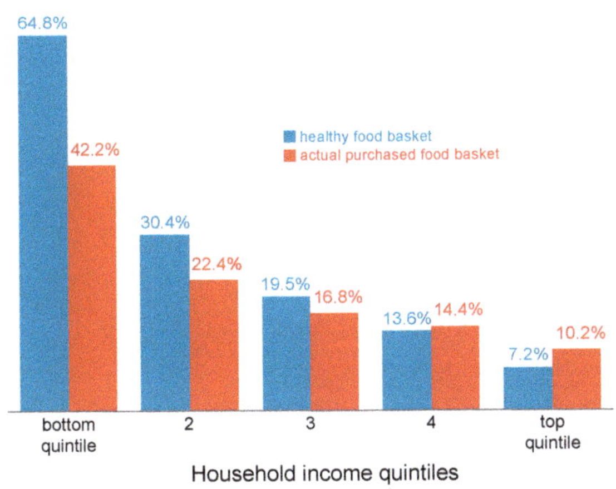

Figure 6. Expenditure on food as percent of household expenditure by income quintile, 2014. This figure is adapted with permission from Ayal Kimhi (2022). Copyright 2022; Copyright Ayal Kimhi.

In the second quintile from the bottom, only 22% of expenses are devoted to food, and this proportion continues to decrease as one moves up the income distribution. If the households in the lowest quintile were to consume a healthy food basket as defined by the Ministry of Health [17], they would have to spend almost two thirds of their total expenses on it, which is unrealistic. (The cost of a healthy food basket is defined as the cheapest way to purchase a food basket assembled according to the recommendations of the Ministry of Health.) Even in the second and third quintiles from the bottom, the cost of a basket of healthy food is higher than the basket of food actually bought.

These findings turn the spotlight towards the prices of healthy food items. Figure 7 shows Israel's rank among 39 OECD countries in terms of the average daily costs of standard food baskets (PPP dollars) in 2017. These costs are based on the cheapest available food items in each country. The three baskets are the standard food basket that meets the caloric needs of the population (2330 kilo-calories), a basket that provides a nutrient-adequate diet, and a basket that provides a healthy diet (A healthy diet provides not only adequate calories but also adequate levels of all essential nutrients and food groups needed for an active and healthy life. The cost of a healthy diet is defined as the cost of the least expensive locally available foods to meet requirements for energy- and food-based dietary guidelines for a representative person within an energy balance at 2,330 kcal/day. The guidelines explicitly recommend food quantities for each food group and provide a wide regional representation [20]). In terms of the prices of a standard food basket, Israel is ranked 16th, above the medial OECD country, indicating that the cost of a standard food basket is higher than in the majority of OECD countries. When comparing the prices of a food basket that provides all the necessary nutrients, Israel is ranked in 32nd place, implying that the prices in Israel are lower than in most OECD countries. This is also the case when comparing the prices of a healthy food basket, as defined by the World Bank, where Israel is ranked 35th, close to the bottom of the distribution. In addition, between 2017 and 2020, the price of the healthy food basket in Israel increased by a little more than 2%, much less than in most developed countries (Figure 8). This implies that the healthy-food-affordability problem of the weaker households is not a problem of local price inflation. Rather, it reflects a global problem.

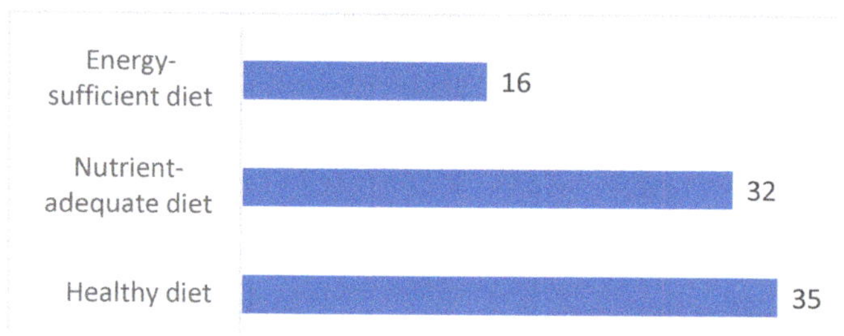

Source: World Bank

Figure 7. Israel's rank in average daily cost of standard food baskets among 39 OECD countries (dollar cost per person per day of cheapest available food), 2017.

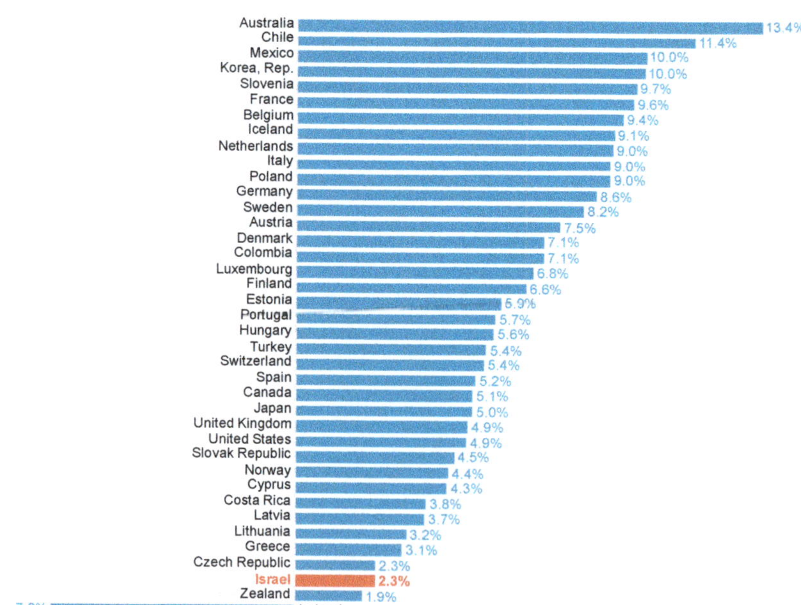

Source: World Bank

Figure 8. Changes in the cost of a healthy food basket between 2017 and 2020.

In this context, it is important to note that for decades Israel has been implementing price controls on products considered basic so that poorer populations will be able to purchase them at reasonable prices. The list of price-controlled products includes bread, salt, dairy products, and eggs [19]. Of the various bread products, price controls are applied to dark bread, white bread, and challah, products that are not considered particularly healthy. In contrast, the healthy food basket includes whole-wheat bread, whose price is double or more than the price of controlled bread, while its production cost is not much higher [21]. The price of salt is also controlled, which may lead to excess consumption with its adverse health consequences. The control over the prices of milk products and eggs is related to the planning policy of the milk- and egg-production system. However, the milk products whose prices are controlled include butter, cream, and hard cheeses which are high in fat, and their consumption in large quantities is not recommended by health professionals. In general, it can be said that the food-price-control policy does not coincide with the promotion of the consumption of a healthy food basket.

Another way to support a healthy diet among vulnerable populations is through direct aid. Civil society organizations operate several programs to supply food directly to needy households, some of which receive government support. The flagship program is the "Food Security Initiative", under which each family receives monthly assistance of three types: a magnetic card worth NIS 250 for use in food chains (without the option of purchasing alcohol or tobacco), fruits and vegetables worth NIS 125, and "dry" food products worth NIS 125. According to the findings of the National Insurance Institute [22], in 2021 there were about 265,000 families in Israel, constituting roughly 8% of the population, that suffered from considerable food insecurity. Of these, only about 11,000 were supported by the food-security initiative.

3.3. Quality and Safety

The quality and safety index includes the subcomponents of dietary diversity, nutritional standards, micronutrient availability, protein quality, and food safety (Table 1). Israel is at the top of the list of developed countries in the areas of protein quality, micronutrient availability, and food safety (Table 1). The areas that slightly lower Israel's position in the field of quality and safety are the low dietary-diversity (24th place) and nutritional standards, which places Israel in the center of the distribution of the OECD countries (The score in the area of nutritional standard is based on four components: Has the government issued guidelines or managed a public program to encourage a balanced diet? Does the government have a national plan to improve nutrition? Does the government require nutritional labeling on food-product packaging? And, does the government monitor the nutritional status of the population?).

3.4. Natural Resources and Resilience

The natural-resources and resilience index includes indices of exposure to climate change, the risk to water quantity and quality, changes in land and aquatic resources, sensitivity to import conditions and natural capital, political commitment to adaptation, and demographic stress (Table 1). Israel is suffering from the depletion of the natural resources needed to produce food, especially land and water. The agricultural sector is gradually losing farmland, especially quality land in the central region, in favor of other land uses such as housing and non-agricultural businesses. Also, the natural water resources that used to be available for agriculture are increasingly either polluted or being taken for other uses, while the desalinated water replacing them is much more expensive (On the other hand, increasingly, treated sewage water is available for agriculture, but this water is not suitable for all crops, and its cost to farmers is at the center of an intense public debate that has not yet been decided). The root of the problem, both in terms of land and water availability, is Israel's rapid population growth and its increasing population density, both of which are unique to developed countries [23].

The problem of the resilience of the food supply in Israel to risks is derived from three main types of risk. One is due to climate change, which is expected to lead to an increase in temperatures, a decrease in precipitation, and above all, increasingly irregular extreme-weather events that impair local food production. The eastern Mediterranean region that Israel belongs to is considered one of the regions where the impact of climate change is expected to be the most severe.

However, when compared to other developed countries, Israel's situation in the area of exposure to climate change is not particularly bad. It is ranked ninth in the OECD in this area; the risk of drought is particularly high, while the risk of flooding is particularly low. The risks of temperature increases and sea levels rising place Israel in the 10th and 14th places in the OECD, respectively (Table 1). Even in a global ranking of nearly 200 countries, Israel is less vulnerable to climate change than most (Figure 9). In contrast, Israel's ranking in the field of readiness to deal with climate change is much lower. Moreover, while Israel's ranking in the field of vulnerability is relatively stable and ranges from 13 to 15 in the period 1995–2020, the ranking in the field of readiness has been steadily declining from the 28th position in 1995 to 41st in 2020. (The vulnerability index represents objective conditions, such as climate change, over which the state has no influence. The readiness index reflects the actions taken by the state in order to deal with the challenges.)

Another type of risk arises from the growing dependence of agricultural production on energy products, most of which are imported to Israel, and whose prices are subject to considerable volatility. The third type of risk lies in the prices of imported food, which are affected by climate change, by supply chain disruptions—such as the one that occurred as a result of the COVID pandemic—and by violent conflicts damaging global food supply, such as Russia's invasion of Ukraine [24]. In this context, it should be noted that Israel imports almost half of its food supply, and if one also adds the import of animal feed, which is necessary for the local production of meat, milk, and eggs, then Israel imports much more than half of its food supply. Although supply sensitivity contributes to only 10% of the general-resilience index, the dependence on food imports places Israel in last place in the OECD in this area.

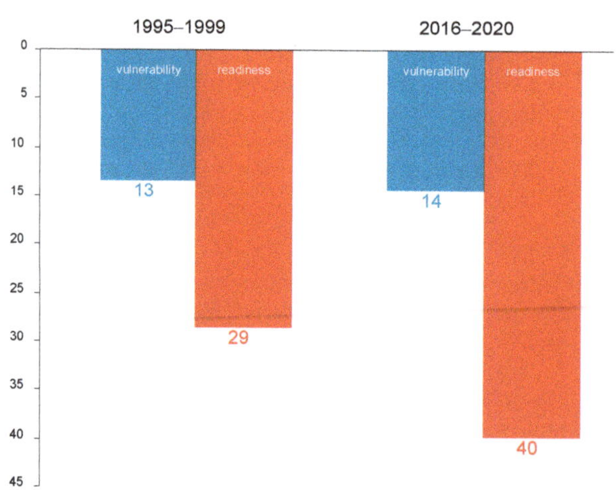

Figure 9. Ranking of Israel in terms of vulnerability and readiness to climate change (the highest the rank, the less vulnerable and the greater the readiness). Source: ND-GAIN [25]. This figure is adapted with permission from Ayal Kimhi (2022). Copyright 2022; Copyright Ayal Kimhi; Source: ND-GAIN (2022).

Figure 10 shows the percentage of self-sufficiency of the main food groups. It can be seen that Israel supplies itself with the lion's share of fresh agricultural produce (the items in the box), while in other food products, which are responsible for a larger share of its caloric supply, the country relies mainly on imports. A calculation based on Figure 10 shows that 56% of Israel's caloric supply relies on imports, and hence, the country's considerable exposure to the risks emanating from global price fluctuations. Regarding the fresh agricultural produce, the high percentage of self-supply is largely rooted in the import restrictions and tariffs. The implementation of import reforms, a political issue that is continuously debated, may lead to a reduction in produce prices in the short term, but will certainly increase the exposure of Israel's food supply to global risks.

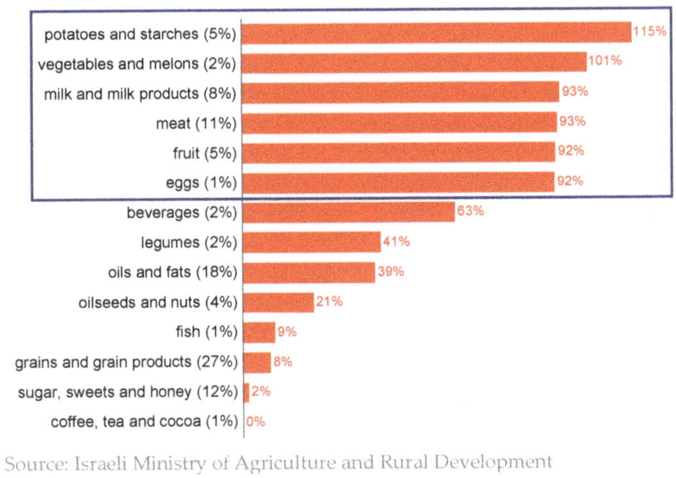

Figure 10. Share of food supplied domestically in 2020 (numbers in parentheses reflect the percent of total energy supply Items with more than 100% sufficiency implies that excess production is exported).

Israel's population is expected to double by 2065 [26]. The rate of increase in agricultural output has been on a downward trend for decades (Figure 1), and given the continued shrinking of cultivated-land areas in favor of urbanization and other land uses such as solar energy, no significant change is expected in this trend if Israel follows a business-as-usual strategy. This implies that the supply of food in Israel will be forced to rely increasingly on imports, with all the attendant risks. In order to preserve the local-production component of food, there will be a need for a significant technological advancement in agriculture that will make it possible to produce more output with fewer inputs, and at a reasonable cost. The next section delves more deeply into the role of agriculture in insuring future food security in Israel.

4. The Role of Agriculture

The growing reliance on food imports in general and the import of fresh agricultural produce in particular, does not diminish, and perhaps even increases, the importance of local agriculture as an important component of Israel's food supply. In recent decades, there has been a noticeable slowdown in the growth of total-factor productivity in global agriculture (An increase in total-factor productivity reflects the increase that would have occurred in the quantity of output had the quantities of the factors of production not changed), both due to a decrease in public funding of research and development, and due to the effects of climate change which are already reflected in damage to crops [27].

In addition, the volatility of prices and supply in world markets is rising as a result of both climate change and crises such as the COVID pandemic and Russia's war in Ukraine [28]. This implies that—especially during this period—it is important for the optimal portfolio of the food basket to include a component of locally produced food, in order to minimize the risk of shortage or a price hike. Toporov et al. [29] concluded that Israel is technically capable of producing food on its own that will satisfy most of the nutritional needs of its residents. However, it is not clear what the cost of such an autarkic policy would be, and whether it is possible to change consumers' feeding habits accordingly, so this conclusion is not particularly relevant. The more relevant question is of which crops can be grown in Israel at a reasonable cost, taking into account both the current alternatives and the future risks? Amdur [30] examined a limited number of agricultural products and found that most of them are imported or can be imported from countries subject to a higher climate risk than Israel, which calls into question the ability to rely on imports of these products in the long term.

Maintaining the local production capacity of fresh agricultural produce requires an adequate income for farmers. Figure 11 shows that, since the 1980s, prices of agricultural inputs have risen faster than the price of output. If agriculture had not benefited from technological advancements and improvements in productivity during this period, it could be concluded that over the years the profitability of agriculture gradually eroded. Improvements in productivity without price changes would have resulted in an increase in profitability. In the absence of a reliable measure of the profitability of agriculture, the fact that many farms have stopped their productive activities over the years suggests that profitability has fallen. At the same time, the gap between the price of food for consumers and the price received by farmers has been widening, so that even during periods when the price of food increased, farmers did not necessarily benefit from this.

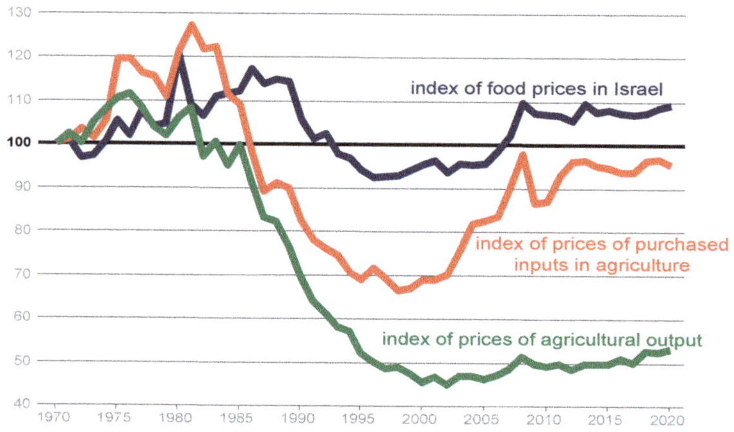

Figure 11. Real price indices (prices discounted by consumer price index) of food and agricultural inputs and output. Source: [31], updated. This figure is adapted with permission from Ayal Kimhi (2022). Copyright 2022; Copyright Ayal Kimhi; Source: Kislev and Zaban (2013), updated.

The decrease in the number of farms is reflected in the decrease in the number of self-employed persons whose main source of income is agriculture (Many farmers also make a living from other gainful activities, either because the income from agriculture is not sufficient or in order to diversify the sources of income due to the large volatility of agricultural income). Figure 12 shows that the number of farmers declined considerably over the years, from about 75,000 in 1960 to about 12,000 in 2020, a decrease of more than 80 percent. The decline was not uniform though. In fact, it can be seen that the rate of decrease in the number of farmers is temporally correlated with the rate of decrease in agricultural output prices, which fell by more than half in real terms during the same period. In the 1970s, output prices stabilized (thanks to the opening of export markets), and at the same time there was a slowdown in the rate of decline in the number of self-employed farmers. The accelerated decline in the prices of agricultural output in the 1980s and 1990s, due to trade liberalization and increased competition, was accompanied by the acceleration of farmers leaving the field. However, the change in the trend of the prices of agricultural output from negative to positive that occurred in the last two decades did not stop the trend of farmers leaving, since the prices of agricultural inputs rose more than the prices of output (Figure 11), suggesting that profitability continued to erode, at least for small and less-productive farms (The agricultural sector consists of large farms (mainly in Kibbutzim) and smaller family farms (mainly in Moshavim). The Kibbutzim (collective farms) did not abandon agriculture even if the number of Kibbutz members involved in agriculture is small. In the family farm sector in the Moshav (cooperative village), on the other hand, a decrease in the number of self-employed farmers implies the exit of the farm from production, and this exit is almost always irreversible. Of course, when small farms exit the sector, other farms have access to more inputs (land and water) and can increase production. This can enhance their own profitability, even when overall profitability in agriculture declines).

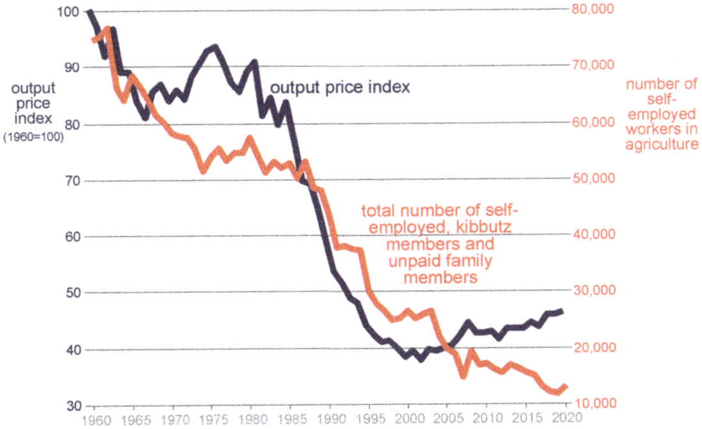

Figure 12. Output-price index and number of self-employed workers in agriculture. This figure is adapted with permission from Ayal Kimhi (2022). Copyright 2022; Copyright Ayal Kimhi; Source: Kislev and Zaban (2013), updated.

The meaning of the decrease in the number of farms at the same time as the increase in agricultural output is that the size (in terms of production capacity) of the average farm increased, and even increased greatly. On the one hand, larger farms can become more efficient by exploiting economies of scale, thereby contributing to food security. On the other hand, the decreasing number of farms may increase the instability of the food supply. For example, it is enough for a number of large farms to stop producing to create an unexpected shortage of certain crops. From this, it follows that it is desirable to create a balance between the need to increase agricultural output and the need to preserve small- and medium-sized family farms. Studies have shown that concentration in a certain industry may have negative consequences not only for competitiveness, but also for innovation in the industry, and on the environment, health, and animal welfare [32].

Increasing investments in agricultural R&D and advanced mechanization may help increase agricultural output and the resilience of the food system. Indeed, one of the components of the agricultural reform proposed by the Ministry of Finance and the Ministry of Agriculture is an increased budget for R&D and capital investments. However, another significant component of the reform is a gradual (over five years) exposure of many crops to competing imports. It is doubtful whether the increase in R&D budgets will bear fruit in such a short period of time, so that only those farms managing to survive will benefit from it, and it is not clear how many of these there will be. Assuming that it will indeed be possible to import certain fruits and vegetables from nearby countries (mainly Turkey, Jordan, and Egypt) at low prices, the local production of those fruits and vegetables will certainly decrease, thereby accelerating the exit, especially among small family farms, from agriculture.

In the long term, the danger inherent in such a scenario is threefold. Farms in those neighboring countries are much less technologically advanced than their Israeli counterparts. Their access to irrigation water during periods of droughts is much more limited, compared to Israeli farms who can use desalinated water. Hence, they are much less prepared to deal with climate change than Israel [33], so the possibility of importing from them at low prices could very well diminish over the years. In addition to this, the political instability in these countries, and the fluctuations in their diplomatic relations with Israel, as recently reflected in the response of their leaders to Israel's war against the Hamas terrorist organization, endanger the regular supply of agricultural products from them. Finally, it is not clear as to what extent the quality of the produce imported from these countries and the environmental and health standards of their farmers can be effectively monitored. The bottom line is that even if the reform will lead to cheaper fruits and vegetables in the short term, it endangers food security in Israel in the long term.

Israel has recently encountered another food-security threat. Israeli farms, and in particular fruit and vegetable farms, rely heavily on the services of foreign workers. Palestinians started working as daily laborers in agriculture, construction, and services since the late 1960s [34], but the worsening security situation during the Palestinian uprising of the late 1980s did not allow them to arrive regularly at work [35], and as a result, the government allowed employers to replace them with foreign workers that came for a five-year period each. The employment of Thai workers in agriculture started in 1993 in small numbers, but those numbers increased considerably in subsequent years [36]. Despite the efforts to limit and even reduce the number of Thai workers in subsequent years, motivated by the populist argument that they take the jobs of local unskilled workers, their numbers remained relatively stable, and agriculture became practically dependent on them [37]. As Figure 13 shows, foreign workers comprise more than a fourth of all agricultural workers in Israel and more than a third of the hired labor force.

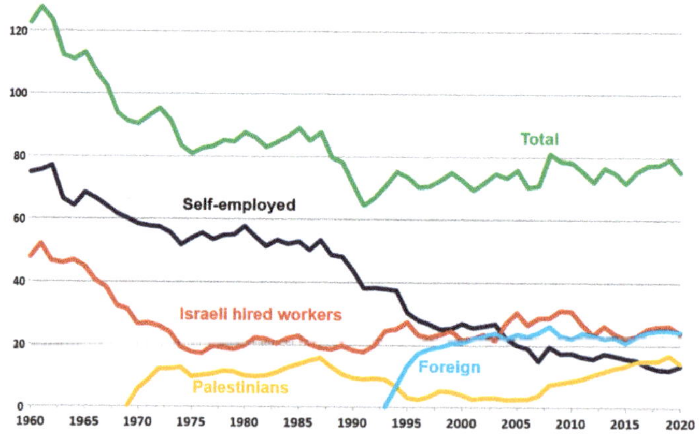

Source: Israel Central Bureau of Statistics

Figure 13. Workers in agriculture.

The atrocities of the Hamas attack on 7 October 2023 and the subsequent war disrupted food security in Israel. On top of the loss of farmland, infrastructure, farm buildings, equipment, and livestock in the agricultural regions that were directly affected, the scores of people murdered, wounded, or kidnapped included foreign agricultural workers. As a result, as many as a third of all Thai workers, not only from the combat zones but from the entire country, left the country immediately. In addition, Palestinians who used to work in agriculture (and other sectors) were banned from entering the country. The immediate and unexpected loss of farm labor implied a threat to food security in both the short and the medium range. Farmers are using volunteers for harvesting this season's crops, but this arrangement cannot continue much longer. Many farmers are hesitating to prepare the fields for next season because they expect that the shortage of labor will sustain. It is likely, then, that the supply of fresh agricultural products will decline in the medium range. The slack could be closed with imports, of course, but those farmers who will stop producing will find it more difficult to resume production in the future because of cash-flow problems. This highlights the importance of a risk-management strategy that will balance local production and imports, so that imports can be augmented when local production is insufficient and local production can be expanded when imports become expensive.

5. Discussion

This paper analyzed the food-security situation in Israel, highlighted the main concerns, and discussed the relevant policy responses. Food security in Israel seems to be satisfactory overall compared to other OECD countries; although, Israel is ranked at the bottom of the list in several areas, most notably resilience to future threats. The intensifying threats of climate change [38], international conflicts, and disruptions in the global supply chain require greater attention from policymakers with an eye to the future. Israel's rapid population growth, which is expected to continue into the foreseeable future, and the slowdown in the rate of growth of its agricultural production over the years, indicate that Israel's reliance on food imports will continue to increase. Importing food exposes Israel even more to global risks, and requires the formulation of a risk-management strategy. Such a strategy must include strengthening local production, especially in products where Israel does not have a significant relative disadvantage. Another threat that requires more attention from policy makers is the combined adverse impact of high and increasing prices of fresh produce and high income inequality that together makes the healthy food basket less affordable to weaker population groups. It should be noted that agricultural adaptation to climate change includes changes in the crop mix that may lead to price changes [39] in a way that can further exacerbate food security among the poor.

The current government's import-exposure policy may contribute to lowering the cost of living and increasing food affordability in the short term, but it increases the country's exposure to outside risk. Specifically, the reduction of tariffs on fruits and vegetables, which, despite the increase in their price in recent years, are still cheaper in Israel than in most developed countries [16], endangers local production capacity and exposes the Israeli consumer to greater future risk. Imports of fruits and vegetables from neighboring countries such as Turkey, Jordan, and Egypt may be attractive under the current conditions. However, the reliance on imports from these countries, which are expected to suffer more from climate change than Israel, may be problematic in the long term, not to mention the inherent risk of geopolitical developments in these countries and their relations with Israel, a risk that becomes more evident since the October 2023 war.

The war has exposed another internal risk though, which is that of labor shortage during periods of conflict. The combination of the internal risks of climate change and labor shortage and the external risks associated with food imports places the durability of the food supply in Israel at a problematic point. However, a risk-management strategy could minimize the threat to future food security. Maintaining the viability of local producers is vital to this goal, and the necessary market reforms should be implemented wisely and gradually so as to minimize uncertainty and suspicion of the farmers [40]. Agricultural R&D investments should be enlarged in order to enhance agricultural productivity and help the sector maintain the supply of fresh produce even as land and other resources continue to shrink. Specific attention should be devoted to developing labor-saving technologies in order to reduce the sector's dependence on foreign workers. Agricultural-insurance programs should be expanded to cover not only weather damages but also those resulting from other events beyond the control of farmers [41,42].

The food security of specific population groups in Israel is affected not only by the availability and price of food, but also by their purchasing power. As Israel is one of the least equal countries in the developed world, it needs a policy focused on its weaker population groups in order to help them obtain a food basket that meets their needs. Moreover, it is necessary to strive for a food basket that will bring these population groups as close as possible to what is defined as a "healthy food basket". Promoting health through healthy food is not only a private interest of each household, but also of society as a whole. In this context, the food-price-control policy, which currently includes many products that are not considered healthy, and the food-aid policy for needy families, which suffers from a rather low budget, must be reconsidered. Many families tend to consume unhealthy food not for economic reasons but due to lack of awareness or lack of understanding of the health consequences [43]. It follows that nutritional education and advocacy (including limiting the advertising of harmful food) should be an integral part of food policy. Economic incentives may also help in cases where education and advocacy are not effective enough. Such incentives may include, for example, taxation of harmful products and price controls of healthy products. These market interventions generally have negative efficiency effects, so their application should be carefully considered subject to a cost–benefit analysis.

There is no shortage of organizations in Israel that deal with food security, but it is necessary for a national body to be established with powers that will enable it to coordinate the activity, supervise the formulation of the strategy, break it down into goals and feasible policy measures, and supervise their implementation. As in the European Union [44], such a body should adopt a holistic approach that deals with all aspects of the food chain (farm to fork), from production in the field, through to processing and marketing, to household consumption. The establishment of such an authority should lead to the stabilization of the policy environment, ease the tension, and restore trust between farmers and the government. This will contribute to the resilience of the agricultural sector and enhance food security.

Food-security policy should be closely linked to the relevant body of research. Future research is particularly required to assess the optimal risk-minimizing portfolio of the local production of fresh agricultural products and their imports.

To summarize, the main conclusions of this paper are as follows:

- Israel's food-security situation is not bad in international comparison, but this is not necessarily sustainable.
- Israel cannot supply all of its food needs, and reliance on imported food is likely to increase.
- As both local production and imports are subject to increasing risks, the formulation of a risk-management strategy is required.
- The farm sector should be stabilized and modernized, with labor-saving R&D investments and a stable policy environment.
- A special authority is needed to establish a long-term food-security strategy.

Funding: This research was funded by the Shoresh Institution for Socioeconomic Research.

Institutional Review Board Statement: Not applicable.

Informed Consent Statement: Not applicable.

Data Availability Statement: The data presented in this study are available on request from the corresponding author. The data are not publicly available due to the fact that it was collected from many different sources.

Acknowledgments: The author acknowledges helpful comments and suggestions by Dan Ben David and other researchers at the Shoresh Institute, by participants in the Workshop on Food Security Strategies for a Changing World at Cornell University, the Israeli Forum for Sustainable Nutrition Workshop on National Food Security in Israel, the 33rd Roundtable Discussion of the Renewing Rural Sector at Tel Aviv University, the Batsheva de Rothschild Workshop on Avoiding the Coming Food Security Crisis, and the 13th annual ESPAnet conference at Haifa University.

Conflicts of Interest: The funders had no role in the design of the study; in the collection, analyses, or interpretation of data; in the writing of the manuscript; or in the decision to publish the results.

References

1. Pinstrup-Andersen, P. Food security: Definition and measurement. *Food Secur.* **2009**, *1*, 5–7. [CrossRef]
2. Ambikapathi, R.; Schneider, K.R.; Davis, B.; Herrero, M.; Winters, P.; Fanzo, J.C. Global food systems transitions have enabled affordable diets but had less favourable outcomes for nutrition, environmental health, inclusion and equity. *Nat. Food* **2022**, *3*, 764–779. [CrossRef] [PubMed]
3. Hawkes, C.; Ambikapathi, R.; Anastasiou, K.; Brock, J.; Castronuovo, L.; Fallon, N.; Malapit, H.; Ndumi, A.; Samuel, F.; Umugwaneza, M.; et al. From food price crisis to an equitable food system. *Lancet* **2022**, *400*, 413–416. [CrossRef] [PubMed]
4. Fuglie, K.; Jelliffe, J.; Morgan, S. Slowing productivity reduces growth in global agricultural output. *Amber Waves Econ. Food Farming Nat. Resour. Rural. Am.* **2021**, 1490–2022-273.
5. Kimhi, A. Heterogeneity, specialization and social cohesion in Israeli Moshav cooperatives. *J. Rural. Coop.* **2009**, *37*, 125–137.
6. Ministry of Industry, Commerce and Tourism. *The Final Report of the Team for Checking the Competitiveness and Prices in the Food and Consumer Products Sectors;* Government of Israel: Jerusalem, Israel, 2012. Available online: https://www.gov.il/he/Departments/publications/reports/food_products_prices_kedmi_report_2012 (accessed on 1 April 2022).
7. Endeweld, M.; Silber, J. Food Security in Israel. In *Food Insecurity in Asia 2017;* Zhou, Z.-Y., Wan, G., Eds.; Asian Development Bank Institute: Tokyo, Japan, 2017; pp. 192–256.
8. Griver, S.; Fischhendler, I. The social construction of food security: The Israeli case. *Food Secur.* **2021**, *13*, 303–1321. [CrossRef]
9. Tal, A. Israeli Agriculture—Innovation and Advancement. In *From Food Scarcity to Surplus: Innovations in Indian, Chinese and Israeli Agriculture;* Gulati, A., Zhou, Y., Huang, J., Tal, A., Juneja, R., Gulati, A., Juneja, R., Huang, J., Eds.; Springer: Singapore, 2021; pp. 299–358.
10. Economist Impact. In *Global Food Security Index 2021;* The Economist Group, 2021; Available online: https://nonews.co/wp-content/uploads/2022/03/GFSI2021.pdf (accessed on 22 November 2022).
11. Alkire, S.; Foster, J. *Designing the Inequality-Adjusted Human Development Index (IHDI);* Human Development Research Paper 2010/28; UNDP: Wadhington, DC, USA, 2010.
12. Adler, D.; Tepper, S.; Tzachor, A. Food security and national resilience in an era of changing climate. In *Environment, Climate and National Security: A New Front for Israel;* Michael, K., Tal, A., Lindenstrauss, G., Bukchin-Peles, S., Hanin, D., Weiss, V., Eds.; Institute for National Security Studies: Tel Aviv, Israel, 2021; Available online: https://www.inss.org.il/he/publication/environment-and-national-security (accessed on 22 November 2022).
13. Kronfeld-Schor, N. *The Climate Crisis and Food Systems;* Lecture at the 33rd Meeting of the Roundtable for the Changing Rural Space; Institute for National Security Studies: Tel Aviv, Israel, 2022.
14. Hamilton, S.F.; Richards, T.J.; Roe, B.E. Food waste: Farms, distributors, retailers, and households. In *Handbook of Agricultural Economics;* Barrett, C.B., Just, D.A., Eds.; Elsevier: Amsterdam, The Netherlands, 2022; Volume 6, pp. 4653–4703.
15. State Comptroller. *Food Loss—Social, Environmental and Economic Consequences;* Annual Report 65c; Government of Israel: Jerusalem, Israel, 2015; pp. 163–235.
16. Ben-David, D.; Kimhi, A. *High Prices in Israel? Beware of International Comparisons;* Shoresh Institution for Socioeconomic Research: Kochav Yair, Israel, 2021.
17. Ministry of Health. *Healthy Food Basket;* Government of Israel: Jerusalem, Israel, 2021. Available online: https://www.gov.il/he/departments/publications/reports/healthy-food-basket-presentation (accessed on 22 November 2022).
18. Ben-David, D.; Kimhi, A. *Israel's Primary Socioeconomic Challenges and Policy Areas Requiring Core Treatment;* Shoresh Institution for Socioeconomic Research: Kochav Yair, Israel, 2017.
19. Azarieva, J.; Arian, B.; Goldsmith, R.; Ginsberg, A.; Milman, R.; Chernichovsky, D. Healthy food basket in Israel. In *State of the Nation Report: Society, Economy and Policy in Israel 2016;* Weiss, A., Ed.; Taub Center for Social Policy Research in Israel: Jerusalem, Israel, 2016; pp. 421–436.
20. FAO; IFAD; UNICEF. *The State of Food Security and Nutrition in the World 2021. Transforming Food Systems for Food Security, Improved Nutrition and Affordable Healthy Diets for All;* FAO: Rome, Italy, 2021.
21. Jetter, K.M.; Cassady, D.L. The availability and cost of healthier food alternatives. *Am. J. Prev. Med.* **2006**, *30*, 38–44. [CrossRef] [PubMed]
22. Endweld, M. *Food Insecurity and Poverty—Israel in the Last Decade;* Lecture at the International Conference on Food Insecurity—The Continuing Pandemic: Toward Sustainable Food Systems; The Israel National Academy of Sciences: Jerusalem, Israel, 2022.
23. Ben-David, D. *Overpopulation and Demography in Israel: Directions, Perceptions, Illustrations and Solutions;* Shoresh Institution for Socioeconomic Research: Kochav Yair, Israel, 2018.
24. Goyal, R.; Steinbach, S. Agricultural commodity markets in the wake of the black sea grain initiative. *Econ. Lett.* **2023**, *231*, 111297. [CrossRef]
25. ND-GAIN, Notre Dame Global Adaptation Initiative. 2022. Available online: https://gain.nd.edu/ (accessed on 22 July 2022).
26. Israel Central Bureau of Statistics. *Population Forecasts for Israel 2015–2065;* Israel Central Bureau of Statistics: Jerusalem, Israel, 2018. Available online: https://www.cbs.gov.il/he/publications/doclib/2019/forcast65/2065/forcast_2016_15.pdf (accessed on 1 April 2022).

27. Barrett, C.B.; Benton, T.; Fanzo, J.; Herrero, M.; Nelson, R.J.; Bageant, E.; Buckler, E.; Cooper, K.; Culotta, I.; Fan, S.; et al. The State of Agri-Food Systems and Agri-Food Value Chains in 2020. In *Socio-Technical Innovation Bundles for Agri-Food Systems Transformation*; Sustainable Development Goals Series; Palgrave Macmillan: Cham, Switzerland, 2022. [CrossRef]
28. Anderson, K. Agriculture in a More Uncertain Global Trade Environment. *Agric. Econ.* **2022**, *53*, 563–579. [CrossRef]
29. Toporov, G.; Greenhut, Z.; Lowengart, A.; Bezek, H.; Zuk-Bar, U.; Kachel, Y. Sustainable nutrition and nutritional security in agriculture in Israel—Quantitative data from crops. *Ecol. Environ.* **2018**, *9*, 18–27.
30. Amdur, L. *The Climate Crisis and Our Plate: How Will Global Climate Change Affect the Food Supply in Israel?* Yesodot Institute: Jerusalem, Israel, 2022.
31. Kislev, Y.; Zaban, S. The Statistical Atlas of Israeli Agriculture. 2013. Available online: https://www.zenovar.com/en/wp-content/uploads/sites/2/2015/06/---------------------.pdf (accessed on 22 July 2023).
32. Crespi, J.M.; MacDonald, J.M. Concentration in food and agricultural markets. In *Handbook of Agricultural Economics*; Barrett, C.B., Just, D.A., Eds.; Elsevier: Amsterdam, The Netherlands, 2022; Volume 6, pp. 4781–4843.
33. Dudu, H.; Cakmak, E.H. Climate change and agriculture: An integrated approach to evaluate economy-wide effects for Turkey. *Clim. Dev.* **2018**, *10*, 275–288. [CrossRef]
34. Bartram, D.V. Foreign Workers in Israel: History and Theory. *Int. Migr. Rev.* **1998**, *32*, 303–325. [CrossRef] [PubMed]
35. Angrist, J.D. Short-Run Demand for Palestinian Labor. *J. Labor Econ.* **1996**, *14*, 425–453. [CrossRef]
36. Miaari, S.H.; Sauer, S.M. The Labor Market Costs of Conflict: Closures, Foreign Workers, and Palestinian Employment and Earnings. *Rev. Econ. Househ.* **2011**, *9*, 129–148. [CrossRef]
37. Kimhi, A. *Are Migrant Agricultural Workers Replacing the Local Workforce?* Discussion Paper No. 10.14; The Center for Agricultural Economic Research: Rehovot, Israel, 2014.
38. Schlesinger, E.Z. Weathering Uncertainty: Climate Change Resilience in Israel's Agriculture. Ph.D. Dissertation, University of Haifa (Israel), Haifa, Israel, 2021.
39. Kan, I.; Reznik, A.; Kaminski, J.; Kimhi, A. The impacts of climate change on cropland allocation, crop production, output prices and social welfare in Israel: A structural econometric framework. *Food Policy* **2023**, *115*, 102311. [CrossRef]
40. FAO. *The Future of Food and Agriculture: Drivers and Triggers for Transformation*; Food and Agriculture Organization of the United Nations: Rome, Italy, 2022.
41. Mahul, O.; Wright, B.D. Designing optimal crop revenue insurance. *Am. J. Agric. Econ.* **2003**, *85*, 580–589. [CrossRef]
42. Smith, V.H.; Glauber, J.W. Agricultural insurance in developed countries: Where have we been and where are we going? *Appl. Econ. Perspect. Policy* **2012**, *34*, 363–390. [CrossRef]
43. Masters, W.A.; Finaret, A.B.; Block, S.A. The economics of malnutrition: Dietary transition and food system transformation. In *Handbook of Agricultural Economics*; Barrett, C.B., Just, D.A., Eds.; Elsevier: Amsterdam, The Netherlands, 2022; Volume 6, pp. 4997–5083.
44. Grant, W.P. *Rethinking Agriculture and Food Policy*; Edward Elgar: Cheltenham, UK, 2022. [CrossRef]

foods

MDPI

Article

Edible Insects: Perceptions of Marketing, Economic, and Social Aspects among Citizens of Different Countries

Raquel P. F. Guiné [1,*], Sofia G. Florença [1], Cristina A. Costa [1], Paula M. R. Correia [1], Luísa Cruz-Lopes [1], Bruno Esteves [1], Manuela Ferreira [2], Anabela Fragata [3], Ana P. Cardoso [3], Sofia Campos [3], Ofélia Anjos [4], Elena Bartkiene [5], Ilija Djekic [6], Irina M. Matran [7], Jelena Čulin [8], Dace Klava [9], Cristina Chuck-Hernández [10], Malgorzata Korzeniowska [11], Nada M. Boustani [12], Maria Papageorgiou [13], Bernardo Prieto Gutiérrez [14], Maša Černelič-Bizjak [15], Emel Damarli [16] and Vanessa Ferreira [17]

1 CERNAS Research Centre, Polytechnic University of Viseu, 3504-510 Viseu, Portugal; sofiaflorenca@outlook.com (S.G.F.); amarocosta@sc.ipv.pt (C.A.C.); paulacorreia@esav.ipv.pt (P.M.R.C.); lvalente@estgv.ipv.pt (L.C.-L.); bruno@estgv.ipv.pt (B.E.)
2 Health Sciences Research Unit: Nursing (UICISA: E), Polytechnic University of Viseu, 3504-510 Viseu, Portugal; mmcferreira@gmail.com
3 CIDEI-IPV Research Centre, Polytechnic University of Viseu, 3504-510 Viseu, Portugal; afragata@estgl.ipv.pt (A.F.); a.p.cardoso@esev.ipv.pt (A.P.C.); sofiamargaridacampos@gmail.com (S.C.)
4 CERNAS Research Centre, Polytechnic University of Castelo Branco, 6001-909 Castelo Branco, Portugal; ofelia@ipcb.pt
5 Department of Food Safety and Quality, Lithuanian University of Health Sciences, 47181 Kaunas, Lithuania; elena.bartkiene@lsmuni.lt
6 Department of Food Safety and Quality Management, Faculty of Agriculture, University of Belgrade, 11000 Belgrade, Serbia; idjekic@agrif.bg.ac.rs
7 Department of Community Nutrition and Food Safety, GEP University MPhScTch of Targu Mures, 540139 Targu Mures, Romania; irina.matran@umfst.ro
8 Maritime Department, University of Zadar, 23000 Zadar, Croatia; jculin@unizd.hr
9 Faculty of Food Technology, Latvia University of Life Sciences and Technologies, LV 3001 Jelgava, Latvia; dace.klava@llu.lv
10 Tecnologico de Monterrey, Institute for Obesity Research, Monterrey 64849, Mexico; cristina.chuck@tec.mx
11 Faculty of Food Science, Wroclaw University of Environmental and Life Sciences, 51-630 Wrocław, Poland; malgorzata.korzeniowska@upwr.edu.pl
12 Faculty of Business and Administration, Saint Joseph University, Beirut 1104 2020, Lebanon; nada.mallahboustany@usj.edu.lb
13 Department of Food Science and Technology, International Hellenic University, 57001 Thessaloniki, Greece; mariapapage@ihu.gr
14 BALAT Research Group, Faculty of Veterinary Medicine, University of León, 24071 León, Spain; bprig@unileon.es
15 Department of Nutritional Counseling–Dietetics, Faculty of Health Science, University of Primorska, 6320 Izola, Slovenia; masa.cernelic@fvz.upr.si
16 Research and Development Center, Altıparmak Food Coop., Çekmeköy, 34782 İstanbul, Turkey; emel.damarli@balparmak.com.tr
17 Department of Nutrition, School of Nursing, UFMG—Federal University of Minas Gerais, Belo Horizonte 30130-100, Brazil; vanessa.nutr@gmail.com
* Correspondence: raquelguine@esav.ipv.pt

Abstract: Because edible insects (EI) have been, in recent years, recommended as a nutritious animal protein food with enormous environmental advantages over other sources of animal protein for human consumption, studies aimed at investigating the consumer perspective have become more prominent. Hence, this study intended to examine the perceptions of participants from different countries about the commercialization and economic and social impacts of edible insects. The study was made using a questionnaire survey, and data were collected in Brazil, Croatia, Greece, Latvia, Lebanon, Lithuania, Mexico, Poland, Portugal, Romania, Serbia, Slovenia, Spain, and Turkey. The final number of received answers was 7222 participants. For the treatment of the results, different statistical techniques were used: factor analysis, internal reliability by Cronbach's alpha, cluster analysis, ANOVA to test differences between groups, and Chi-square tests. The results obtained confirmed the validity of the scale, constituted by 12 out of the 14 items initially considered, distributed by 4 factors: the first related to the economic impact of EIs, the second related to the motivation for consumption of EIs, the third related to the places of purchase of EIs, and the fourth corresponding to a question presented to the participants as a false statement. A cluster analysis allowed identifying three clusters, with significant differences between them according to all the sociodemographic variables tested. Also, it was found that the participants expressed an exceptionally high level of agreement with aspects such as the difficulty in finding EIs on sale, knowledge acting as a strong motivator for EI consumption, and the role of personalities and influencers in increasing the will to consume EIs. Finally, practically all sociodemographic variables were found to be significantly associated with perceptions (country, sex, education, living environment, and income), but not age. In conclusion, the perceptions about EI commercialization were investigated and revealed differences among samples originating from different countries. Moreover, the sociodemographic characteristics of the participants were found to be strongly associated with their perceptions.

Keywords: insect consumption; economy; social impact; questionnaire survey; marketing

1. Introduction

The United Nations (UN) estimates a continuing rise in the world population, which is a high challenge for sustainable development. The UN released in 2015 the Sustainable Development Goals (SDG) as a global movement to end poverty, protect the planet, and ensure that, by 2030, all people enjoy peace and prosperity [1].

Food security is at risk, and there is rising apprehension regarding the availability of the food required to meet the needs of the growing population [2,3]. Food production impacts the planet's limited resources, such as farmland and freshwater, both when it comes to the production of vegetable or animal foods [4–9], but fisheries also have a huge impact on aquatic ecosystems [10,11]. Climate change also poses serious risks to the sustainable development of the planet and for food production. Factors such as lack of water, extreme weather conditions, droughts, floods, fires impacting forests and agricultural fields, and insect pests infecting agricultural crops cause important annual losses of food and are expected to be more severe in the future [12,13]. Therefore, high motivation has been observed globally among not only the scientific community, but also society in general and some political organizations to minimise the impacts of global warming and unsustainable development [1,14–16].

Alternative food sources, providing high nutritive value with more efficient use of resources, are a trend that aims to help sustainable development and efficient food providence. Edible insects (EI), among other sustainable trends like cultured meat or algae-based foods, constitute a way to meet these challenges, providing high protein contents more sustainably compared to conventional sources of protein for human consumption, like beef or pork protein [17,18]. For this reason, EIs have been suggested by the FAO—Food and Agriculture Organization of the United Nations [19] as part of a global strategy to achieve sustainable food security.

Although insects have been part of the cultural culinary traditions among entomophagic communities in some parts of the world, their acceptance by consumers in most Western countries is faced with reluctance, suspicion, and even disgust or neophobia. Owing to these differences, the consumption and commercialization of EIs or insect-based foods are highly variable according not only to geographic factors, but also to other types of motivation [20–22]. Owing to these differences in food patterns among countries, the production and commercialization of insects are also very different among countries. While in some countries, insects are easily sold at all commercialization points, from street markets to other shops, in European countries, only some insects are allowed to be commercialized, as approved by the European Food Safety Authority (EFSA) under the Novel Food Regulation, and they are available in fewer selling points [23]. In the European Markets, edible insects and their derived ingredients are allowed to be marketed only after they have been authorized. This marketing authorisation is provided by a number of agents involved, namely the European Commission (EC), the safety evaluation of the novel food by the European Food Safety Authority (EFSA), and a favourable vote given by the EU Member States (MS) [24]. Under these restrictions, the farming of edible insects in European countries is also limited by EU regulations [25]. In contrast, in countries where entomophagy is part of the traditional food culture, the production and commercialization of edible insects are facilitated. The commercialization of edible insects was presented as a livelihood opportunity for local communities in the Lake Victoria basin [26]. The species that are produced and marketed also differ according to culture. For example, grasshoppers and small hemipterans known as 'jumiles' have been reported as the two main insects sold in the markets of the state of Morelos, Mexico [27].

Because edible insects' consumption is highly variable among countries, and as a consequence, their commercialization is also differentiated, it is of interest to investigate how edible insects are marketed in different social–cultural environments, as well as to understand the role of consumers in these aspects. Hence, the aim of this study was to investigate the perceptions of participants from different countries about the commercialization and economic and social impacts of edible insects. This could be of particular importance to better understand how markets function and adapt selling strategies to increase consumers' buying intentions. Also, in the ambit of this study, possible differences according to sociodemographic variables, such as country, sex, age, education, living environment, and income, were analysed. Finally, the scale items were analysed using a factor analysis to assess consistency.

2. Materials and Methods

2.1. Instrument and Data Collection

The data used in this study were collected by employing a questionnaire survey implemented by the team of the project EiSuFood, which was designed to investigate consumer issues related to EIs [28]. The Ethics Committee of the Polytechnic University of Viseu approved this questionnaire survey with reference 45/SUB/2021, and all ethical principles were strictly followed when designing the questionnaire and collecting the data, including those of the Declaration of Helsinki. Only adult citizens (18 to 88 years) who had expressed their informed consent were allowed access to the questionnaire to participate in the survey, and they were allowed to quit at any time without sending the answers if they wanted to. The questionnaire was answered online (on the Google Docs platform) and the invitation was made through an internet link shared by email or social media.

Table A1 in Appendix A presents the 14 questions included in the questionnaire that focused on the topics addressed in this particular study, namely two dimensions, Economic and Social aspects and Commercialization and Marketing. To measure the participants' perceptions, they were asked to express their agreement with each item on a central five-point Likert scale as follows: 1 = strongly disagree, 2 = disagree, 3 = no opinion, 4 = agree, and 5 = strongly agree [29]. The questionnaire was prepared in the ambit of the EISuFood Project and validated through Structural Equation Modelling [28].

Although it is a convenience sample, some hint of the possible minimum sample size was calculated as indicative. For this, some assumptions were considered: Targeting 50% of the adult population; Confidence interval = 90%; Z score = 1.645; Power of the test = 5% (minimum acceptable probability of preventing type II error = 0.05) [30,31]. Based on these assumptions, for an infinite population, the number of responses stabilizes at 271. So, considering all countries involved, even for those with the highest population, a minimum of 271 participants should be guaranteed [32,33].

The data collection was carried out in 14 countries (Brazil, Croatia, Greece, Latvia, Lebanon, Lithuania, Mexico, Poland, Portugal, Romania, Serbia, Slovenia, Spain, and Turkey) in the second semester of the year 2021, and a total of 7222 participations were recorded. The initial target in each country was to obtain a minimum of 300 responses.

2.2. Data Analysis

To evaluate the scale items, a Factor Analysis (FA) with the extraction methods of a Principal Components Analysis and Varimax rotation were applied. To fix the number of components to extract, the criterion used was Eigenvalues greater than 1. The Kaiser–Meyer–Olkin (KMO) measure to assess the adequacy of the sample was computed. Also the, Bartlett's test was made to analyse the correlations between the variables studied [34]. For reference, values of KMO over 0.5 are acceptable, and the higher they are, the more suited the data are for the use of FA techniques [35]. Factor loadings with absolute values lower than 0.4 were excluded [36,37]. Cronbach's alpha (α) was used to determine the internal consistency of each of the factors extracted [34,38]. The reference values for alpha considered were: acceptable consistency—$0.5 < \alpha < 0.7$, good consistency—$0.7 \leq \alpha < 0.8$, and very good consistency—$0.8 \leq \alpha$ [39–41].

Complementing FA, a cluster analysis was also performed based on the factors obtained through FA. For this, hierarchical clustering was applied with the Ward method and considering the Squared Euclidean distance to measure the intervals. For hierarchical clustering, the variables used were the four factors, and the results of the coefficients in the agglomeration schedule allowed fixing the number of clusters in three. Then, the same Ward method was run again but with fixing the number of clusters in three, and the solution was saved in the database for a posterior analysis of the clusters.

To evaluate differences between groups, ANOVA—analysis of variance was used, coupled with the post hoc test of Tukey, to identify differences between the means among groups. Also, chi-square tests were utilized when analysing differences between clusters based on sociodemographic variables. The coefficients of Cramer's V were considered to evaluate the strength of the associations between the categorical variables tested. The reference values considered for the Cramer's V coefficient were: $V \approx 0.1$—weak association, $V \approx 0.3$—moderate association, and $V \approx 0.5$ or more—strong association [42].

The level of significance considered was 5%, and the software used in all statistical analyses was IBM® SPSS Statistics software (Version 28).

3. Results

3.1. Sample Characterization

Figure 1 presents the distribution of the 7222 participants by country, indicating a higher participation from Mexico (n = 1139) and a lower participation from Turkey and Latvia (n = 296 and n = 300, respectively).

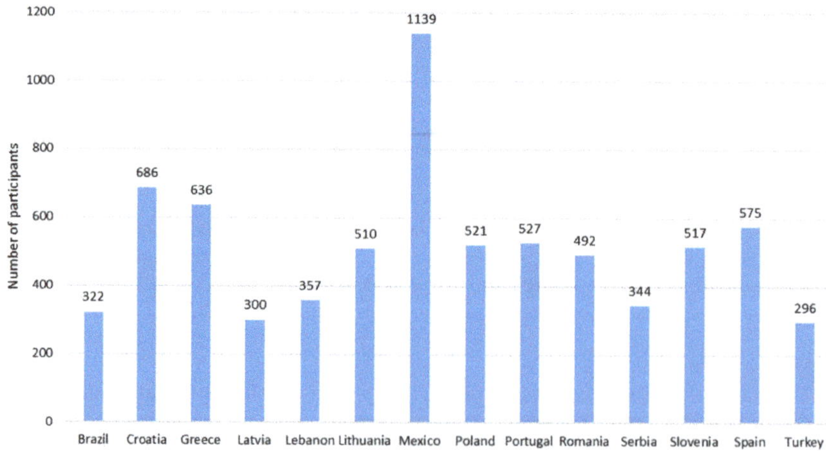

Figure 1. Distributions of the participants by country (N = 7222).

Figure 2 shows the distribution of the participants according to age, sex, education, living environment, and income. The majority of the participants were female (63%), and the age class most represented was young adults (aged between 18 and 30 years, which represented 47% of the sample). With respect to education, 36% did not have a university graduation, while 32% had a university degree and 32% had post-graduate studies. A high percentage of participants resided in urban environments (66%), and most participants (38%) had an income equal to the average income in their own countries.

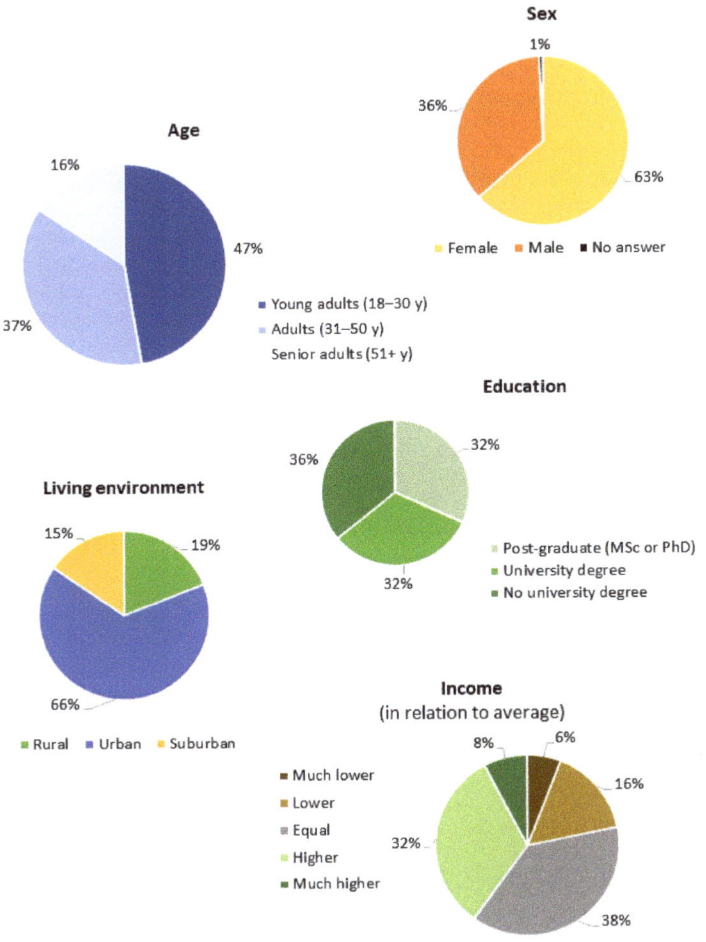

Figure 2. Sociodemographic characteristics of the sample according to age, sex, education, living environment, and income.

3.2. Analysis of the Scale Items with Factor Analysis

Bartlett's test was significant ($p < 0.0005$), thus leading to the rejection of the null hypothesis "H0: The correlation matrix is equal to the identity matrix", and, in fact, in the correlation matrix, there were some associations between the variables, with two values above 0.5 (r = 0.588 between V1 and V2, r = 0.502 between V1 and V5) (The list of variables is listed as the items in Appendix A).

According to the reference values for the KMO [35], the obtained value of 0.836 can be considered as good. All the values of the Measure of Sampling Adequacy (MSA) were higher than 0.5, based on the anti-image matrix, confirming that all the variables could be included in the analysis. The lowest MSA was 0.592 for variable V8, and the highest was 0.894 for variable V13.

The FA with the PCA and Varimax rotation extracted four factors that explained 56.31% of the variance. Practically all communalities were higher than 0.5, except for variables V8 (0.467) and V14 (0.463), so these two variables were found as not suitable to be included in the analysis. For this reason, a second FA was run without variables V8 and V14.

For the second FA, Bartlet's test was again significant ($p < 0.0005$) and the value of KMO increased to 0.848, and the values of MSA also increased, resulting in a minimum of 0.712 (for V7) and a maximum of 0.899 (for V13). This solution extracted four factors again, explaining 62.0% of the variance. This percentage indicates that only about 56% of the variance in the data could be explained by this solution, and the rest of the variance is determined to other conditions. In addition, the factor solution was obtained assuming the criteria of getting eigenvalues greater than 1, as indicated in Materials and Methods,, which is usually the criteria for choosing the factors to be included in an analysis. All communalities were higher than 0.5, the highest value being 0.847 (variable V3 has 84.7% of its variance explained by the solution), and the lowest being 0.534 (53.4% VE for variable V10). The results obtained are shown in Table 1, and they include the validation by Cronbach's alpha (α), measuring the internal consistency of each factor [34]. The distribution of the items by the factors resulted in one factor (F1) that contained five variables related to the economic impact of EIs, another factor (F2) that included four variables related to the motivation for the consumption of EIs, a third factor (F3) that included two variables, both related to the places of purchase of EIs, and a fourth factor (F4) that included only one variable, which was the only item that was given as a false statement. The internal consistency of factors F1 and F2 was good ($\alpha = 0.785$ and $\alpha = 0.702$, respectively), while for factor F3, the value of alpha was acceptable ($\alpha = 0.560$).

Table 1. The final solution obtained through factor analysis.

Factor	%VE [1]	Item	Loading	Alpha [2]
F1	22.6%	V1. Insect production can contribute to increase the income of families in low-income areas	0.692	0.785
		V2. Insects provide protein foods at cheap prices	0.694	
		V4. Presently, the Asia–Pacific and Latin America areas account for more than half of the edible insects' market	0.692	
		V5. In some countries insect farming is becoming a key factor to fight against rural poverty	0.774	
		V6. The income generated from insects can be affected by market fluctuations in price derived from availability	0.677	
F2	17.0%	V10. The level of knowledge influences the willingness to purchase insect food	0.557	0.702
		V11. Price is among the motivations to consume insect foods	0.789	
		V12. The consumption of insects and derived foods depends on availability	0.777	
		V13. Personalities/influencers can lead people to consume insects	0.547	
F3	12.3%	V7. Edible insects are difficult to find on sale in street markets	0.820	0.560
		V8. Edible insects are on sale only in specialized shops	0.777	
F4	9.5%	V3. The market for edible insects is expected to decline in the future	0.920	N.A. [3]

[1] VE = Variance explained. [2] Cronbach's alpha. [3] N.A. = not applicable, single item in the factor.

Figure 3 presents a graphical representation of the factors in the rotated space, clearly showing the separation of the four factors. Factor F1 was the most relevant, including five variables (V1, V2, V4, and V6), that, together, explain 22.6% of the total variance. Factor F2 explains 17.0% of the total variance and contains four variables (V10, V11, V12, and V13). Factor F3 contains only two variables (V7 and V8) and explains 12.3% of variance, while the last factor, F4, explains 9.5% of the variance and included only one variable (V3).

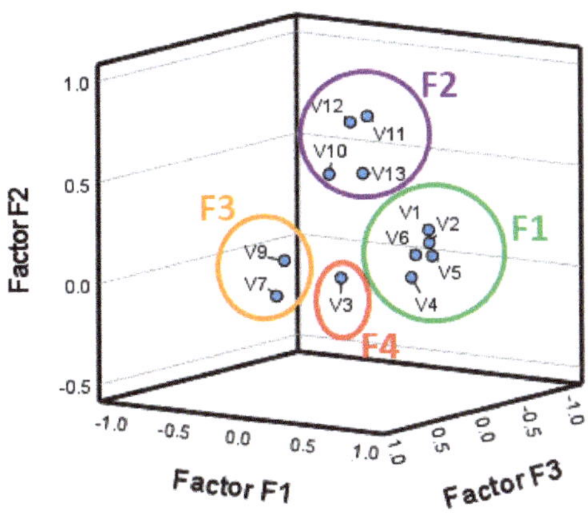

Figure 3. Factor Analysis component plot for solution obtained with Varimax rotation.

3.3. Cluster Analysis

Based on the factorial structure obtained, a cluster analysis was performed, allowing for distinguishing three clusters to group the cases. These groups of participants were characterized according to the sociodemographic variables as presented in Table 2. Significant differences at the level of 5% ($p < 0.05$) were observed between the clusters for the three sociodemographic variables considered (sex, age, and education), but the associations were weak, as indicated by the low values of the Cramer's coefficients. A higher percentage of female participants (42.6%) fit into cluster C1, while male participants were mostly in cluster C2 (38.9%). With respect to age, while cluster C2 had more young adults (40.8%), cluster C1 had more adults (44.1%) and senior adults (44.2%). In terms of education level, cluster C1 contained a higher percentage of the participants with the highest level of instruction (48.5%).

Table 2. Association between cluster membership and sociodemographic variables.

Variables	Groups	Cluster C1 (%)	Cluster C2 (%)	Cluster C3 (%)	Total (%)
Sex	Female	42.6	36.2	21.2	100.0
(p [1] < 0.001;	Male	37.4	38.9	23.8	100.0
V [2] = 0.038)	No answer	31.3	45.8	22.9	100.0
Age group	Young adults (18–30 y)	36.5	40.8	22.6	100.0
(p [1] < 0.001;	Adults (31–50 y)	44.1	34.5	21.4	100.0
V [2] = 0.058)	Senior adults (51 y or over)	44.2	33.2	22.5	100.0
Education level	Post graduate (MSc ot PhD)	48.5	32.3	19.2	100.0
(p [1] < 0.001;	University degree	36.6	40.3	23.0	100.0
V [2] = 0.078)	No University degree	37.2	38.8	24.0	100.0

[1] Significance of the Chi-square test. [2] Cramer's coefficient.

Table 3 presents the results of the cross-tabulation between the cluster membership and the geographical variables, country and living environment. In both cases, the Chi-square test indicated significant differences ($p < 0.05$), although the association was moderate for variable country (V = 0.320), while being low for living environment (V = 0.030). The results reveal that cluster C1 contains more Greek (59.9%), Latvian (66.0%), Lithuanian (45.3%), Polish (62.7%), Portuguese (58.3%), Romanian (47.4%), and Spanish (51.5%) participants, while in cluster C2 predominate the Brazilians (49.7%), Croatians (42.8%), Lebanese (73.7%), Mexicans (41.6%), Serbians (60.2%), Slovenians (42.9%), and Turkish (58.8%) participants. Regarding living environment, also presented in Table 3, the results highlight similar distributions of rural and suburban participants between clusters C1 and C2, while those living in urban environments are clearly more representative in cluster C1 (41.1%).

Table 3. Association between cluster membership and demographic variables.

Variables	Groups	Cluster C1 (%)	Cluster C2 (%)	Cluster C3 (%)	Total (%)
Country (p [1] < 0.001; V [2] = 0.320)	Brazil	37.9	49.7	12.4	100.0
	Croatia	24.2	42.8	33.0	100.0
	Greece	59.9	4.4	35.7	100.0
	Latvia	66.0	15.3	18.7	100.0
	Lebanon	22.4	73.7	3.9	100.0
	Lithuania	45.3	41.0	13.7	100.0
	Mexico	26.6	41.6	31.8	100.0
	Poland	62.7	13.8	23.5	100.0
	Portugal	58.3	38.5	3.2	100.0
	Romania	47.4	45.3	7.3	100.0
	Serbia	11.9	60.2	27.9	100.0
	Slovenia	27.3	42.9	29.8	100.0
	Spain	51.5	20.0	28.5	100.0
	Turkey	36.5	58.8	4.7	100.0
Living environment (p [1] = 0.012; V [2] = 0.030)	Rural	40.3	40.4	19.3	100.0
	Urban	41.1	36.3	22.5	100.0
	Suburban	38.7	37.4	23.9	100.0

[1] Significance of the Chi-square test. [2] Cramer's coefficient.

Table 4 contains the results for the cross-tabulation between income and cluster membership, revealing that clusters C1 and C2 have similar percentages of participants in practically all income categories, except for the high-income participants, who are present in cluster C1 in a higher percentage (42.8%). In this case, the differences were significant p ($p < 0.05$), but the association was low (V = 0.048).

Table 4. Association between cluster membership and economic variables.

Variables	Groups	Cluster C1 (%)	Cluster C2 (%)	Cluster C3 (%)	Total (%)
Income (p [1] < 0.001; V [2] = 0.048)	Much lower than average	37.0	35.8	27.1	100.0
	Lower than average	38.2	37.0	24.8	100.0
	Equal to average	39.8	37.7	22.5	100.0
	Higher than average	42.8	35.8	21.5	100.0
	Much higher than average	42.9	41.9	15.2	100.0

[1] Significance of the Chi-square test. [2] Cramer's coefficient.

3.4. Perceptions of the Participants

Table 5 presents the results obtained for each of the items isolated, and they show a high level of agreement towards all items except for items V3, V8, and V14. As previously seen, questions V8 and V14 were not considered to be consistent enough to be part of the scale, and in relation to question V3, this was included, but as an isolated factor. Question V3 corresponds to a false statement, whereas questions V8 and V14 may be strongly dependent on the origin of the participants, since, in the sample, most participants were from non-insect-eating countries, but there were also many participants from countries where Eis are part of the culinary culture, such as, for example, Mexico, in some of the country's regions.

Table 5. Scores attributed for each item, considering the global sample.

Variable	Items	% of Answers				
		Level of Agreement [1]				
		1	2	3	4	5
V1	Insect production can contribute to increase the income of families in low-income areas	5.7	9.9	32.3	37.4	14.7
V2	Insects provide protein foods at cheap prices	5.3	11.2	33.7	34.3	15.4
V3	The market for edible insects is expected to decline in the future	17.5	33.4	37.2	8.6	3.3
V4	Presently, the Asia–Pacific and Latin America areas account for more than half of the edible insects' market	2.9	5.3	42.1	35.8	13.8
V5	In some countries insect farming is becoming a key factor to fight against rural poverty	3.5	6.7	46.8	33.0	9.9
V6	The income generated from insects can be affected by market fluctuations in price derived from availability	3.5	7.3	47.8	31.3	10.2
V7	Edible insects are difficult to find on sale in street markets	5.9	10.9	21.2	35.2	26.9
V8	Edible insects are easy to find on sale in supermarkets	34.5	32.6	20.5	8.0	4.4
V9	Edible insects are on sale only in specialized shops	4.6	8.9	28.9	39.6	18.0
V10	Level of knowledge influences the willingness to purchase insect food	4.8	7.6	21.7	41.0	24.9
V11	Price is among the motivations to consume insect foods	10.4	17.7	33.8	27.6	10.4
V12	The consumption of insects and derived foods depends on availability	6.2	14.4	28.4	37.4	13.6
V13	Personalities/influencers can lead people to consume insects	5.5	7.7	20.5	40.9	25.4
V14	Insect consumption is independent of marketing campaigns	18.9	31.5	27.9	15.1	6.6

[1] Values on a scale from 1 to 5, for which: 1 = strongly disagree, 2 = disagree, 3 = no opinion, 4 = agree, and 5 = strongly agree.

Items with a very high percentage of agreement or strong agreement included, for example, V7, V10, and V13, dealing with, respectively, difficulty in finding EIs on sale, the knowledge acting as a strong motivator for EI consumption, and the role of personalities and influencers in increasing the will to consume EIs.

For the items considered problematic, V3, V8, and V14, as well as for those that gathered the highest level of agreement globally, V7, V10, and V13, the responses obtained from the participants from each country are shown in Figures 4 and 5. For the other variables that were not so relevant, similar results are presented in Appendix B, in table format.

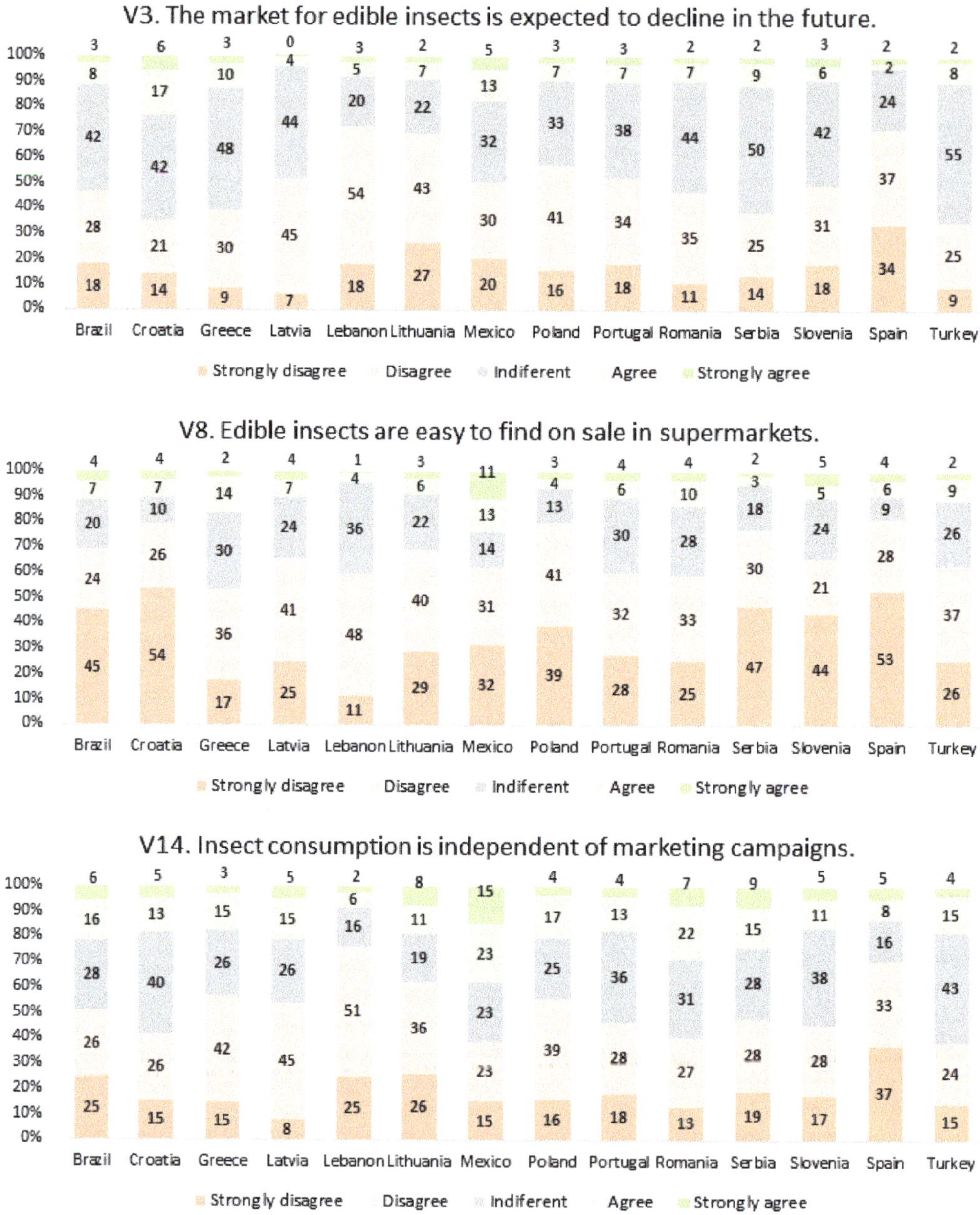

Figure 4. Responses obtained to items V3, V8, and V14, separated by country.

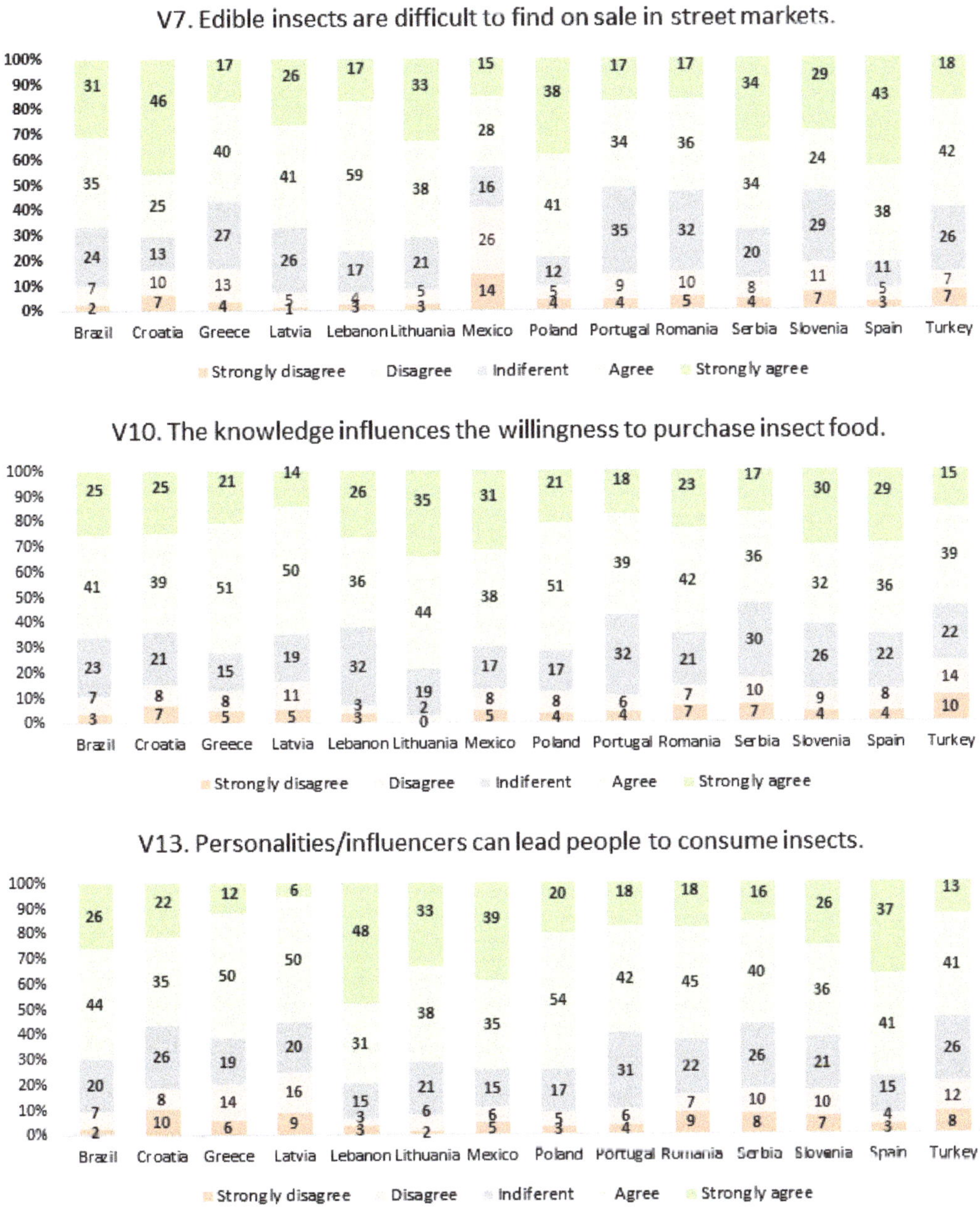

Figure 5. Responses obtained to items V7, V10, and V13, separated by country.

Regarding the results in Figure 4, country differences are visible for all three items. For V3, the false item separated by the FA into one factor, F4, the highest percentage of people who did not manifest an opinion was found for Turkey (55% of indifferent score), and the lowest for Lebanon (20%). Since this statement corresponds to a false affirmation, responding disagreement scores corresponded to a correct perception. Therefore, higher percentages of disagreement scores (1 = strongly disagree and 2 = disagree, altogether) were found for Lebanon (18 + 54 = 72%), Spain (24 + 37 = 71%), and Lithuania (27 + 43 = 70%), revealing a predominance of correct perception in these countries. The two other variables considered to be problematic were items V8 and V14, which were excluded from the FA for not being consistent enough. The results in Figure 4 show that, for V8, the only country where a slightly higher percentage of agreement scores was found (4 = agree and 5 = strongly agree, jointly) was Mexico (13 + 11 = 24%), showing that, in some regions of Mexico, the commercialization of insects is a regular practice. Also, the results for item V14 show a similar trend, with a predominance of positive answers for Mexican participants, with 38% (23 + 15 = 38%) believing that consumption of EIs does not depend on marketing campaigns.

In Figure 5, it presents the results obtained for items that gathered the highest level of agreement globally. For item V7, the results show that EIs are generally difficult to find on sale in street markets, particularly in Lebanon (17 + 59 = 79%), Poland (38 + 41 = 79%), and Spain (38 + 43 = 81%). In what concerns the influence of knowledge in the purchase of EIs (item V10), highest agreement was found for Lithuania (35 + 44 = 79%), Greece (21 + 51 = 72%), and Poland (21 + 51 = 72%). Finally, the results for item V13 revealed that, in Lebanon (31 + 48 = 79%) and Spain (41 + 37 = 78%), the highest percentages of agreement were found for the influence of personalities and influencers in encouraging the consumption of EIs.

A mean score was calculated for each participant considering the responses given to all the items, and those values were then tested for differences according to sociodemographic variables, country, age, sex, education, living environment, and income. The results presented in Table 6 show significant differences according to country, sex, education, living environment, and income. On the other hand, age was not found to be associated with the perceptions, since no significant differences were observed between age groups. Nevertheless, in this case, the value of significance was marginal ($p = 0.053$), just slightly above the level of significance considered in the analyses ($p < 0.05$).

Table 6. Perceptions according to sociodemographic variables.

Variables	Groups	Perceptions Mean ± s.d. [1]	Significance [2]
Country	Brazil	3.24 ± 0.44 [bcd]	$p < 0.001$
	Croatia	3.16 ± 0.50 [abc]	
	Greece	3.18 ± 0.49 [abc]	
	Latvia	3.12 ± 0.31 [a]	
	Lebanon	3.47 ± 0.49 [f]	
	Lithuania	3.26 ± 0.33 [cd]	
	Mexico	3.42 ± 0.5 [ef]	
	Poland	3.31 ± 0.39 [de]	
	Portugal	3.18 ± 0.49 [abc]	
	Romania	3.23 ± 0.51 [bcd]	
	Serbia	3.13 ± 0.57 [ab]	
	Slovenia	3.22 ± 0.48 [abcd]	
	Spain	3.25 ± 0.43 [cd]	
	Turkey	3.14 ± 0.55 [ab]	
Sex	Female	3.24 ± 0.48 [a]	$p = 0.014$
	Male	3.28 ± 0.50 [ab]	
	No answer	3.31 ± 0.57 [b]	
Age class	Young adults (18–30 y)	3.26 ± 0.50 [a]	$p = 0.053$
	Adults (31–50 y)	3.26 ± 0.47 [a]	
	Senior adults (51+ y)	3.22 ± 0.47 [a]	
Education	Post-graduate (MSc or PhD)	3.25 ± 0.44 [ab]	$p < 0.001$
	Completed university degree	3.28 ± 0.49 [b]	
	No university degree	3.23 ± 0.52 [a]	

Table 6. *Cont.*

Variables	Groups	Perceptions Mean ± s.d. [1]	Significance [2]
Living environment	Rural	3.22 ± 0.51 [a]	$p = 0.007$
	Urban	3.26 ± 0.49 [ab]	
	Suburban	3.28 ± 0.46 [b]	
Household income in relation to average	Much lower	3.25 ± 0.58 [ab]	$p = 0.009$
	Lower	3.28 ± 0.49 [b]	
	Equal	3.23 ± 0.49 [a]	
	Higher	3.27 ± 0.45 [b]	
	Much higher	3.26 ± 0.51 [ab]	
Global sample		3.25 ± 0.49	

[1] Values on a scale from 1 to 5, for which: 1 = strongly disagree, 2 = disagree, 3 = no opinion, 4 = agree, and 5 = strongly agree. [2] ANOVA with post hoc test of Tukey at a level of significance of 5% ($p < 0.05$). Mean values with different superscripts are significantly different.

4. Discussion

Edible insects may be capable of providing economic benefits, since they seem to be a more sustainable and environmentally friendly nutrient source than other animals and have a high feed-to-meat conversion efficiency [43]. More than two billion people globally routinely consume insects [44]. In 130 countries, 3071 ethnic groups consume over 2086 insect species, with the African, Australian, Asian, and South American continents traditionally being the most entomophagous regions [43,45].

The results of our research revealed differences between countries, sex, education, living environment, and income, and therefore, these sociodemographic characteristics were found to significantly influence perceptions about EIs. Different cultures accept the consumption of EIs as food to different degrees. Acceptance can be influenced by complex themes such as regional values, trends in health and sustainability, and cultural attitudes towards food. For example, northern Europeans tend to have a more positive view of insect food than central Europeans [46]. According to a study developed in Italy by Laureati et al. [47], age, gender, awareness of the topic, and food neophobia were found to be the most influential factors in this regard for Italian consumers. The readiness to accept insects was stronger among males than females and was stronger among younger consumers than among older consumers.

Food and food service industries can tailor their marketing efforts to better meet the specific needs of different consumer segments, taking into account these sociodemographic characteristics, which allows companies to develop strategies to achieve EI consumer development and adapt pricing and promotion [48].

Despite the numerous benefits of EI consumption, consumer acceptance still remains one of the obstacles to their utilization as a protein food source, especially in developed countries where insects are viewed in disgust by the majority of the population [49]

According to Wendin and Nyberg [50], even though sustainability factors are important, they are seldom the main reasons influencing insect consumption. Instead, a complexity of emotional factors, such as disgust and neophobia, as well as familiar tastes, textures, and contexts, were found to have a major influence. In addition, exposure and positive tasting experiences have been identified as important factors for increasing acceptance.

Our study revealed that knowledge is a strong motivator for the consumption of EIs, so we believe that a strategy should be developed by food manufacturers to educate consumers on the nutritional benefits, changing the negative perceptions (thus increasing consumer willingness to eat insects). Both Mancini et al. [51] and Barton et al. [52] also proposee that education paired with informative tasting sessions could be one strategy to reduce the rejection of and disgust for the use of insects as food. Jones [53] suggested that another strategy could be teaching some subjects in the classroom about sustainable food choices, but also in all areas of the school, for example, in the school canteen.

Huis et al. [44] compiled in their study some strategies to convince customers to consume of EIs: (1) emphasising that insects are nutritionally adequate; (2) incorporating them in an unrecognisable form in familiar products; (3) making insect products delicious; (4) giving people a taste experience; (5) marketing insect-based products by taste; (6) providing detailed information about the insect product, taking into consideration that sustainability may not be the most convincing factor; (7) using celebrities to promote the product; (8) targeting specific groups such as sensation seekers or children; and (9) devising market strategies, such as using stylistic images and choosing supermarkets for retailing.

According to Ventanas et al. [54], marketing strategies should emphasise positive emotions (i.e., adventurous) to persuade potential consumers.

Megido et al. [55] observed that people are more willing to consume EIs when presented in other food forms, e.g., cookies, energy bars, burgers, and sandwich spreads, among others, and, according to a study conducted by Meyer-Rochow et al. [56], insects processed into flour or pastes have a greater chance of being accepted when insect images are not displayed on the packaging.

Our study also revealed that a major concern of consumers was finding EIs on sale in supermarkets and that they are mainly found in specialized shops. In a study conducted in Thailand by Phonthanukitithaworn et al. [57], marketing strategies are crucial and should include attractive advertising and promotion strategies, as well as accessible distribution channels; promotional tools such as buy-one-get-one-free could stimulate insect consumption. In a study conducted by Alemu et al. [58], in Kenya, they concluded that affirmative recommendations are particularly important for processed EIs, and consumers prefer to buy this type of product in kiosks or supermarkets than at local marketplaces.

Our study also revealed that personalities/influencers can lead people to consume insects. Also, according to a study conducted by Lensvelt and Steenbekkers [59], if the information about Eis is provided by scientists, close relatives, the government, or persons using the product, it may influence EI consumption. Moreover, Park et al. [60] found that males in the USA could be influenced by using advertisements for EI, featuring actors/actresses and athletes, while for females, it was only the first group.

5. Conclusions and Limitations

The results of the scale analysis revealed a structure composed of four factors that explained 62% of the variance, one factor gathering the items related to the economic impact of EIs, another factor including the items related to the motivation for consumption of EIs, a third factor joining together two items related to the places of purchase of EIs, and a fourth factor containing a single item, which was the question presented to the participants as a false statement. The FA excluded two items, one related to the easiness of finding EIs on sale in supermarkets and another about the influence of marketing campaigns on the consumption of EIs. The internal consistency of factors F1 and F2 was found to be good, while for factor F3, it was acceptable. Factor F4 is a single item, therefore, no internal reliability applies to this case. Taking into consideration the obtained results, it was concluded that the questionnaire items' factorial structure containing four factors was found to be appropriate for assessing the perceptions of consumers about the economic and social impact of EIs in multiple countries.

The results further showed that aspects related to difficulty in finding EIs on sale, knowledge acting as a strong motivator for EI consumption, and the role of personalities and influencers in increasing the will to consume EIs, received by the participants a high level of agreement, indicating strong perceptions about these issues.

Finally, sociodemographic variables like country, sex, education, living environment, and income were found to significantly influence the perceptions of the participants in the study, while age was not found to be significantly related to the perceptions.

This work, although providing valuable insight, has some limitations that can, to some extent, limit the applicability of the results. One of them is related to the uneven distribution of the participants between countries. While the populations in the different countries are highly variable in size, the number of obtained responses was not directly proportional, and this resulted from the difficulty in recruiting participants to respond to the questionnaire, especially in some of the countries. For example, Brazil is a very large country, and the number of obtained responses was rather low compared to Croatia, a much smaller country. One other limitation was related to the uneven distribution of the participants by the different classes according to the sociodemographic variables considered, namely much more female participants, who usually are more predisposed to answer questionnaires, more participants living in urban environments, a high percentage of post-graduate participants, and a high representativeness also of young adults up to 30 years. These dissimilarities can introduce some bias into the interpretation of the results.

Author Contributions: Conceptualisation, R.P.F.G., C.A.C., P.M.R.C., M.F., A.P.C., S.C. and O.A.; methodology, R.P.F.G.; software, R.P.F.G.; validation, R.P.F.G.; formal analysis, R.P.F.G.; investigation, R.P.F.G., C.A.C., P.M.R.C., S.G.F., M.F., A.P.C., S.C., O.A., C.C.-H., V.F., M.P., I.M.M., B.P.G., M.K., M.Č.-B., E.B., J.Č., N.M.B., I.D., D.K. and E.D.; resources, R.P.F.G., C.A.C., P.M.R.C., M.F., A.P.C., S.C. and O.A.; data curation, R.P.F.G.; writing—original draft preparation, R.P.F.G., A.F. and S.G.F.; writing—review and editing, R.P.F.G., L.C.-L., B.E., M.Č.-B., N.M.B., M.P. and J.Č.; supervision, R.P.F.G.; project administration, R.P.F.G.; funding acquisition, R.P.F.G., C.A.C., P.M.R.C., M.F., A.P.C. and S.C. All authors have read and agreed to the published version of the manuscript.

Funding: This work is funded by National Funds through the FCT—Foundation for Science and Technology, I.P., within the scope of the projects Ref. UIDB/00681/2020 (CERNAS), UIDB/05507/2020 (CIDEI) and UIDB/00742/2020 (UICISA:E) research unit.

Institutional Review Board Statement: This research followed string ethical procedures, ensuring respect for all ethical standards. Also, the guidelines of the Declaration of Helsinki were respected during investigation and data handling. This research to be carried out by questionnaire survey received approval on 25 May 2020 by the ethics committee of Polytechnic University of Viseu (Reference No. 45/SUB/2021).

Informed Consent Statement: Informed consent was obtained from all subjects involved in the study.

Data Availability Statement: Data may be requested from the corresponding author.

Acknowledgments: This work was supported by the FCT—Foundation for Science and Technology, I.P. Furthermore, we would like to thank the CERNAS, CIDEI, and UCISA:E Research Centers and the Polytechnic University of Viseu for their support. This research was developed in the ambit of the project "EISuFood—Edible Insects as Sustainable Food", with Reference CERNAS-IPV/2020/003.

Conflicts of Interest: Author Emel Damarli was employed by the company Research and Development Center, Altıparmak Food Coop., Çekmeköy, 34782 İstanbul, Turkey. This author's contribution is investigation. The remaining authors declare that the research was conducted in the absence of any commercial or financial relationships that could be construed as a potential conflict of interest.

Appendix A

The items of the questionnaire used to investigate the Economic and Social aspects (items 1 to 6) as well as the Commercialization and Marketing aspects of EIs (items 7 to 14) are shown in Table A1.

Table A1. List of items in the questionnaire (Scale: 1 = Strongly disagree; 2 = Disagree; 3 = No opinion; 4 = Agree; and 5 = Strongly agree).

Variable	Items	Level of Agreement				
		1	2	3	4	5
1	Insect production can contribute to increase the income of families in low-income areas					
2	Insects provide protein foods at cheap prices					

Table A1. *Cont.*

Variable	Items	Level of Agreement				
		1	2	3	4	5
3	The market for edible insects is expected to decline in the future					
4	Presently, the Asia–Pacific and Latin America areas account for more than half of the edible insects' market					
5	In some countries insect farming is becoming a key factor to fight against rural poverty					
6	The income generated from insects can be affected by market fluctuations in price derived from availability					
7	Edible insects are difficult to find on sale in street markets					
8	Edible insects are easy to find on sale in supermarkets					
9	Edible insects are on sale only in specialized shops					
10	The level of knowledge influences the willingness to purchase insect food					
11	Price is among the motivations to consume insect foods					
12	The consumption of insects and derived foods depends on availability					
13	Personalities/influencers can lead people to consume insects					
14	Insect consumption is independent of marketing campaigns					

Appendix B

The results of the percentages of the scale scores obtained for variables V1, V2, V4, V5, V6, V9, V11, and V12, according to country are shown in Table A2.

Table A2. Percentage of participants' responses according to country, for the less relevant variables.

Variable [1]	Score [2]	Percentage of Responses													
		Country [3]													
		BR	HR	GR	LV	LB	LT	MX	PL	PT	RO	RS	SI	ES	TR
V1	1	4.3	12.1	6.8	5.3	3.1	7.5	3.1	3.5	4.7	6.3	7.3	4.8	5.0	7.4
	2	7.8	15.3	11.2	19.7	3.1	14.5	5.8	5.0	6.5	10.0	13.1	8.7	7.8	20.6
	3	37.0	32.7	37.3	44.0	19.3	29.0	17.9	31.3	39.1	33.9	41.0	40.4	36.9	32.8
	4	35.1	29.9	40.3	27.7	54.9	33.3	39.0	48.3	39.1	43.5	31.7	31.5	34.1	33.1
	5	15.8	10.1	4.6	3.3	19.6	15.7	34.2	11.9	10.6	6.3	7.0	14.5	16.2	6.1
V2	1	2.8	12.0	4.6	4.7	2.8	8.2	3.4	2.5	4.6	5.9	3.8	3.9	6.4	7.8
	2	8.1	18.7	11.0	13.3	2.8	19.4	11.5	6.7	8.0	10.4	9.3	9.9	10.8	10.1
	3	41.6	43.1	42.3	39.0	13.7	29.6	19.8	25.8	41.4	40.7	49.1	40.6	29.4	32.4
	4	32.0	20.8	35.4	40.0	47.6	28.4	34.2	50.2	35.1	37.4	28.5	28.0	32.9	41.6
	5	15.5	5.4	6.8	3.0	33.1	14.3	31.2	14.8	11.0	5.7	9.3	17.6	20.5	8.1
V4	1	1.9	5.8	2.4	1.0	2.0	6.5	2.2	1.3	2.3	3.5	3.8	1.4	1.7	3.7
	2	9.6	7.9	4.6	2.7	2.5	11.8	5.1	2.5	4.2	3.5	7.6	4.4	4.0	4.1
	3	54.0	42.2	38.8	51.0	29.1	38.4	39.9	43.3	49.9	47.8	44.2	49.9	26.3	47.6
	4	28.6	28.9	45.1	41.7	48.5	30.2	33.0	42.7	34.9	37.8	32.8	28.0	40.5	33.8
	5	5.9	15.2	9.1	3.7	17.9	13.1	19.8	10.2	8.7	7.5	11.6	16.2	27.5	10.8
V5	1	1.9	6.3	3.9	1.3	2.2	5.5	3.2	1.9	4.4	3.3	5.2	1.5	3.1	4.1
	2	6.5	11.4	5.0	5.7	3.1	15.3	5.3	2.7	5.5	6.9	9.6	6.4	6.3	3.7
	3	54.0	42.9	48.9	68.0	36.7	39.0	41.7	53.7	49.1	41.1	50.9	50.9	45.6	51.7
	4	29.2	33.1	37.9	23.7	39.8	28.0	33.5	35.0	33.0	40.2	27.9	28.6	31.5	34.5
	5	8.4	6.4	4.2	1.3	18.2	12.2	16.3	6.7	8.0	8.5	6.4	12.6	13.6	6.1

Table A2. *Cont.*

Variable [1]	Score [2]	Percentage of Responses													
		Country [3]													
		BR	HR	GR	LV	LB	LT	MX	PL	PT	RO	RS	SI	ES	TR
V6	1	1.2	6.9	2.8	2.7	2.2	5.3	3.1	1.9	3.0	4.5	5.2	2.1	2.6	4.4
	2	10.6	8.6	8.5	6.0	3.9	12.9	7.1	2.5	5.1	7.5	7.8	7.0	7.3	5.7
	3	57.5	54.5	47.5	60.0	37.3	29.2	34.9	57.5	57.1	48.4	55.2	53.4	43.5	58.8
	4	23.6	23.9	35.5	31.0	37.8	32.2	36.9	33.1	28.5	35.4	24.4	26.3	32.5	26.4
	5	7.1	6.1	5.7	0.3	18.8	20.4	18.0	5.0	6.3	4.3	7.3	11.2	14.1	4.7
V9	1	3.4	4.7	3.5	1.0	2.5	2.5	10.2	4.0	3.4	3.5	6.1	3.3	3.1	5.4
	2	6.5	5.1	11.2	3.7	2.5	3.3	19.1	5.8	7.2	5.5	12.2	7.7	9.7	8.4
	3	35.7	23.0	23.9	43.7	23.8	25.3	18.2	29.0	43.3	33.5	34.6	37.3	24.0	39.2
	4	36.0	38.0	50.5	42.7	54.9	34.5	36.1	49.2	36.8	41.5	30.8	32.7	38.4	33.4
	5	18.3	29.2	11.0	9.0	16.2	34.3	16.4	11.9	9.3	16.1	16.3	19.0	24.7	13.5
V11	1	10.6	14.7	12.3	10.0	4.2	16.1	7.0	5.4	6.6	9.6	9.9	13.0	15.8	10.8
	2	12.4	16.5	27.2	31.7	3.6	28.2	14.6	15.4	14.0	16.7	19.5	16.8	16.2	18.2
	3	44.1	34.8	25.8	31.3	31.1	29.8	23.6	37.3	52.9	38.2	37.8	33.1	39.8	26.4
	4	25.8	24.2	28.1	25.0	41.7	16.7	34.9	34.0	20.9	28.5	25.6	25.9	19.3	34.1
	5	7.1	9.8	6.6	2.0	19.3	9.2	19.9	7.9	5.5	7.1	7.3	11.2	8.9	10.5
V12	1	7.5	8.7	9.0	5.7	2.8	5.7	3.2	4.2	6.8	5.9	7.0	7.9	6.3	8.8
	2	12.7	13.8	25.6	20.0	7.0	12.0	9.7	15.6	13.7	12.0	14.0	15.3	16.9	16.2
	3	35.4	31.3	32.9	22.7	27.2	19.0	18.0	20.2	40.4	29.9	35.8	36.4	30.4	32.1
	4	33.2	34.5	28.8	46.3	46.2	39.2	42.3	52.1	32.8	41.7	32.6	29.2	30.8	34.5
	5	11.2	11.5	3.8	5.3	16.8	24.1	26.8	7.9	6.3	10.6	10.8	11.2	15.7	8.4

[1] V1 = Insect production can contribute to increase the income of families in low-income areas, V2 = Insects provide protein foods at cheap prices, V4 = Presently, the Asia–Pacific and Latin America areas account for more than half of the edible insects' market, V5 = In some countries insect farming is becoming a key factor to fight against rural poverty, V6 = The income generated from insects can be affected by market fluctuations in price derived from availability, V9 = Edible insects are on sale only in specialized shops, V11 = Price is among the motivations to consume insect foods, and V12 = The consumption of insects and derived foods depends on availability. [2] Scale: 1 = Strongly disagree; 2 = Disagree; 3 = No opinion; 4 = Agree; and 5 = Strongly agree. [3] BR = Brazil, HR = Croatia, GR = Greece, LV = Latvia, LB = Lebanon, LT = Lithuania, MX = Mexico, PL = Poland, PT = Portugal, RO = Romania, RS = Serbia, SI = Slovenia, ES = Spain, and TR = Turkey.

References

1. UN Sustainable Development Goals | United Nations Development Programme. Available online: https://www.undp.org/sustainable-development-goals (accessed on 22 August 2023).
2. Carrillo-Álvarez, E. Perspective: Food and Nutrition Insecurity in Europe: Challenges and Opportunities for Dietitians. *Adv. Nutr.* **2023**, *14*, 995–1004. [CrossRef] [PubMed]
3. Mejía, G.; Aránguiz, R.; Espejo-Díaz, J.A.; Granados-Rivera, D.; Mejía-Argueta, C. Can Street Markets Be a Sustainable Strategy to Mitigate Food Insecurity in Emerging Countries? Insights from a Competitive Facility Location Model. *Socio-Econ. Plan. Sci.* **2023**, *86*, 101456. [CrossRef]
4. Fusco, G.; Campobasso, F.; Laureti, L.; Frittelli, M.; Valente, D.; Petrosillo, I. The Environmental Impact of Agriculture: An Instrument to Support Public Policy. *Ecol. Indic.* **2023**, *147*, 109961. [CrossRef]
5. Gill, M.; Garnsworthy, P.C.; Wilkinson, J.M. Review: More Effective Linkages between Science and Policy Are Needed to Minimize the Negative Environmental Impacts of Livestock Production. *Animal* **2021**, *15*, 100291. [CrossRef] [PubMed]
6. Li, J.; Yang, W.; Liu, L.; Liu, X.; Qiu, F.; Ma, X. Development and Environmental Impacts of China's Livestock and Poultry Breeding. *J. Clean. Prod.* **2022**, *371*, 133586. [CrossRef]
7. Meier, M.S.; Stoessel, F.; Jungbluth, N.; Juraske, R.; Schader, C.; Stolze, M. Environmental Impacts of Organic and Conventional Agricultural Products–Are the Differences Captured by Life Cycle Assessment? *J. Environ. Manag.* **2015**, *149*, 193–208. [CrossRef]
8. Skinner, J.A.; Lewis, K.A.; Bardon, K.S.; Tucker, P.; Catt, J.A.; Chambers, B.J. An Overview of the Environmental Impact of Agriculture in the U.K. *J. Environ. Manag.* **1997**, *50*, 111–128. [CrossRef]
9. van Selm, B.; Hijbeek, R.; van Ittersum, M.K.; van Hal, O.; van Middelaar, C.E.; de Boer, I.J.M. Recoupling Livestock and Feed Production in the Netherlands to Reduce Environmental Impacts. *Sci. Total Environ.* **2023**, *899*, 165540. [CrossRef]
10. Babcock, E.A.; Tewfik, A.; Burns-Perez, V. Fish Community and Single-Species Indicators Provide Evidence of Unsustainable Practices in a Multi-Gear Reef Fishery. *Fish. Res.* **2018**, *208*, 70–85. [CrossRef]
11. Steadman, D.; Appleby, T.; Hawkins, J. Minimising Unsustainable Yield: Ten Failing European Fisheries. *Mar. Policy* **2014**, *48*, 192–201. [CrossRef]
12. Mirzabaev, A.; Bezner Kerr, R.; Hasegawa, T.; Pradhan, P.; Wreford, A.; Cristina Tirado von der Pahlen, M.; Gurney-Smith, H. Severe Climate Change Risks to Food Security and Nutrition. *Clim. Risk Manag.* **2023**, *39*, 100473. [CrossRef]

13. Subedi, B.; Poudel, A.; Aryal, S. The Impact of Climate Change on Insect Pest Biology and Ecology: Implications for Pest Management Strategies, Crop Production, and Food Security. *J. Agric. Food Res.* **2023**, *14*, 100733. [CrossRef]
14. Ferranti, P. 1.18-Politics for Food Security and Climate Changes. In *Sustainable Food Science—A Comprehensive Approach*; Ferranti, P., Ed.; Elsevier: Oxford, UK, 2023; pp. 176–182. ISBN 978-0-12-824166-0.
15. Roy, P.; Pal, S.C.; Chakrabortty, R.; Chowdhuri, I.; Saha, A.; Shit, M. Climate Change and Groundwater Overdraft Impacts on Agricultural Drought in India: Vulnerability Assessment, Food Security Measures and Policy Recommendation. *Sci. Total Environ.* **2022**, *849*, 157850. [CrossRef] [PubMed]
16. Zembe, A.; Nemakonde, L.D.; Chipangura, P. A Policy Coherence Framework for Food Security, Climate Change Adaptation and Disaster Risk Reduction in South Africa. *Int. J. Disaster Risk Reduct.* **2023**, *95*, 103877. [CrossRef]
17. Borges, M.M.; da Costa, D.V.; Trombete, F.M.; Câmara, A.K.F.I. Edible Insects as a Sustainable Alternative to Food Products: An Insight into Quality Aspects of Reformulated Bakery and Meat Products. *Curr. Opin. Food Sci.* **2022**, *46*, 100864. [CrossRef]
18. Guiné, R.P.F.; Correia, P.; Coelho, C.; Costa, C.A. The Role of Edible Insects to Mitigate Challenges for Sustainability. *Open Agric.* **2021**, *6*, 24–36. [CrossRef]
19. *FAO Looking at Edible Insects from a Food Safety Perspective-Challenges and Opportunities for the Sector*; FAO-Food and Agriculture Organization of the United Nations: Rome, Italy, 2021.
20. Bisconsin-Júnior, A.; Rodrigues, H.; Behrens, J.H.; da Silva, M.A.A.P.; Mariutti, L.R.B. "Food Made with Edible Insects": Exploring the Social Representation of Entomophagy Where It Is Unfamiliar. *Appetite* **2022**, *173*, 106001. [CrossRef]
21. Florença, S.G.; Guiné, R.P.F.; Gonçalves, F.J.A.; Barroca, M.J.; Ferreira, M.; Costa, C.A.; Correia, P.M.R.; Cardoso, A.P.; Campos, S.; Anjos, O.; et al. The Motivations for Consumption of Edible Insects: A Systematic Review. *Foods* **2022**, *11*, 3643. [CrossRef] [PubMed]
22. Mishyna, M.; Chen, J.; Benjamin, O. Sensory Attributes of Edible Insects and Insect-Based Foods–Future Outlooks for Enhancing Consumer Appeal. *Trends Food Sci. Technol.* **2020**, *95*, 141–148. [CrossRef]
23. EFSA Edible Insects: The Science of Novel Food Evaluations. Available online: https://www.efsa.europa.eu/en/news/edible-insects-science-novel-food-evaluations (accessed on 1 November 2023).
24. IPIFF. *Commercialisation of Edible Insects in the EU*; International Platform of Insects for Food & Feed: Brussels, Belgium, 2023.
25. Żuk-Gołaszewska, K.; Gałęcki, R.; Obremski, K.; Smetana, S.; Figiel, S.; Gołaszewski, J. Edible Insect Farming in the Context of the EU Regulations and Marketing—An Overview. *Insects* **2022**, *13*, 446. [CrossRef] [PubMed]
26. Odongo, W.; Okia, C.A.; Nalika, N.; Nzabamwita, P.H.; Ndimubandi, J.; Nyeko, P. Marketing of Edible Insects in Lake Victoria Basin: The Case of Uganda and Burundi. *J. Insects Food Feed.* **2018**, *4*, 285–293. [CrossRef]
27. Pino Moreno, J.M.; Reyes-Prado, H. Commerce of Edible Insects in the State of Morelos, Mexico. *J. Insect Sci.* **2020**, *20*, 19. [CrossRef] [PubMed]
28. Guiné, R.P.F.; Duarte, J.; Chuck-Hernández, C.; Boustani, N.M.; Djekic, I.; Bartkiene, E.; Sarić, M.M.; Papageorgiou, M.; Korzeniowska, M.; Combarros-Fuertes, P.; et al. Validation of the Scale Knowledge and Perceptions about Edible Insects through Structural Equation Modelling. *Sustainability* **2023**, *15*, 2992. [CrossRef]
29. Likert, R. A Technique for the Measurement of Attitudes. *Arch. Psychol.* **1932**, *22*, 5–55.
30. Triola, M.F.; Flores, V.R.L.F. *Instrodução àEstatística*, 12th ed.; LTC: Rio de Janeiro, Brasil, 2017.
31. Levin, J.; Fox, J.A. *Estatística Para Ciências Humanas*, 9th ed.; Pearson: Rio de Janeiro, Brasil, 2004.
32. Cochran, W.G. *Sampling Techniques*, 3rd ed.; John Wiley & Sons: New York, NY, USA, 1977.
33. Levine, D.M.; Stephan, D.F.; Krehbiel, T.C.; Berenson, M.L. *Estatistica Teoria E Aplicacoes Usando O Microsoft Excel Em Portugues*, 5th ed.; LTC: Rio de Janeiro, Brasil, 2008; ISBN 978-85-216-1634-4.
34. Broen, M.P.G.; Moonen, A.J.H.; Kuijf, M.L.; Dujardin, K.; Marsh, L.; Richard, I.H.; Starkstein, S.E.; Martinez–Martin, P.; Leentjens, A.F.G. Factor Analysis of the Hamilton Depression Rating Scale in Parkinson's Disease. *Park. Relat. Disord.* **2015**, *21*, 142–146. [CrossRef]
35. Kaiser, H.F.; Rice, J. Little Jiffy, Mark Iv. *Educ. Psychol. Meas.* **1974**, *34*, 111–117. [CrossRef]
36. Stevens, J.P. *Applied Multivariate Statistics for the Social Sciences*, 5th ed.; Routledge: New York, NY, USA, 2009; ISBN 978-0-8058-5903-4.
37. Rohm, A.J.; Swaminathan, V. A Typology of Online Shoppers Based on Shopping Motivations. *J. Bus. Res.* **2004**, *57*, 748–757. [CrossRef]
38. Tanaka, K.; Akechi, T.; Okuyama, T.; Nishiwaki, Y.; Uchitomi, Y. Development and Validation of the Cancer Dyspnoea Scale: A Multidimensional, Brief, Self-Rating Scale. *Br. J. Cancer* **2000**, *82*, 800–805. [CrossRef]
39. Hair, J.F.H.; Black, W.C.; Babin, B.J.; Anderson, R.E. *Multivariate Data Analysis*, 7th ed.; Prentice Hall: Upper Saddle River, NJ, USA, 2009; ISBN 978-0-13-813263-7.
40. Maroco, J.; Garcia-Marques, T. Qual a fiabilidade do alfa de Cronbach? Questões antigas e soluções modernas? *Laboratório De Psicol.* **2006**, *4*, 65–90. [CrossRef]
41. Davis, F.B. *Educational Measurements Their Interpretation*; Wadsworth Pub. Co.: Belmont, CA, USA, 1964.
42. Witten, R.; Witte, J. *Statistics*, 9th ed.; Wiley: Hoboken, NJ, USA, 2009.
43. Lange, K.W.; Nakamura, Y. Edible Insects as Future Food: Chances and Challenges. *J. Future Foods* **2021**, *1*, 38–46. [CrossRef]

44. Huis, A.V.; Itterbeeck, J.A.; Klunder, H.; Mertens, E.; Halloran, A.; Muir, G.; Vantomme, P. *Edible Insects-Future Prospects for Food and Feed Security Edible Insects Future Prospects for Food and Feed Security*; FAO-Food and Agriculture Organization of the United Nations: Rome, Italy, 2013.
45. Ramos-Elorduy, J. Anthropo-Entomophagy: Cultures, Evolution and Sustainability. *Entomol. Res.* **2009**, *39*, 271–288. [CrossRef]
46. Piha, S.; Pohjanheimo, T.; Lähteenmäki-Uutela, A.; Křečková, Z.; Otterbring, T. The Effects of Consumer Knowledge on the Willingness to Buy Insect Food: An Exploratory Cross-Regional Study in Northern and Central Europe. *Food Qual. Prefer.* **2018**, *70*, 1–10. [CrossRef]
47. Laureati, M.; Proserpio, C.; Jucker, C.; Savoldelli, S. New Sustainable Protein Sources: Consumers' Willingness to Adopt Insects as Feed and Food. *Ital. J. Food Sci.* **2016**, *28*, 652–668. [CrossRef]
48. Lee, J.; Bae, S.J. Attributes of Insect Food Acceptance: Identifying Key Factors with Consumer Market Segmentation. *Int. J. Gastron. Food Sci.* **2023**, *32*, 100702. [CrossRef]
49. Imathiu, S. Benefits and Food Safety Concerns Associated with Consumption of Edible Insects. *NFS J.* **2020**, *18*, 1–11. [CrossRef]
50. Wendin, K.M.; Nyberg, M.E. Factors Influencing Consumer Perception and Acceptability of Insect-Based Foods. *Curr. Opin. Food Sci.* **2021**, *40*, 67–71. [CrossRef]
51. Mancini, S.; Moruzzo, R.; Riccioli, F.; Paci, G. European Consumers' Readiness to Adopt Insects as Food. A Review. *Food Res. Int.* **2019**, *122*, 661–678. [CrossRef]
52. Barton, A.; Richardson, C.D.; McSweeney, M.B. Consumer Attitudes toward Entomophagy before and after Evaluating Cricket (Acheta Domesticus)-Based Protein Powders. *J. Food Sci.* **2020**, *85*, 781–788. [CrossRef]
53. Jones, V. 'Just Don't Tell Them What's in It': Ethics, Edible Insects and Sustainable Food Choice in Schools. *Br. Educ. Res. J.* **2020**, *46*, 894–908. [CrossRef]
54. Ventanas, S.; González-Mohino, A.; Olegario, L.S.; Estévez, M. Newbie Consumers Try Pizzas in Which Bacon Is Replaced by *Tenebrio molitor* L. Larvae: Not as Healthy as Expected and Not as Terrible as They Thought. *Int. J. Gastron. Food Sci.* **2022**, *29*, 100553. [CrossRef]
55. Megido, R.C.; Sablon, L.; Geuens, M.; Brostaux, Y.; Alabi, T.; Blecker, C.; Drugmand, D.; Haubruge, É.; Francis, F. Edible Insects Acceptance by Belgian Consumers: Promising Attitude for Entomophagy Development. *J. Sens. Stud.* **2014**, *29*, 14–20. [CrossRef]
56. Meyer-Rochow, V.B.; Hakko, H. Can Edible Grasshoppers and Silkworm Pupae Be Tasted by Humans When Prevented to See and Smell These Insects? *J. Asia-Pac. Entomol.* **2018**, *21*, 616–619. [CrossRef]
57. Phonthanukitithaworn, C.; Sae-eaw, A.; Tang, H.; Chatsakulpanya, P.; Wang, W.; Ketkaew, C. Marketing Strategies and Acceptance of Edible Insects Among Thai and Chinese Young Adult Consumers. *J. Int. Food Agribus. Mark.* **2023**, *35*, 154–182. [CrossRef]
58. Alemu, M.H.; Olsen, S.B.; Vedel, S.E.; Pambo, K.O.; Owino, V.O. Combining Product Attributes with Recommendation and Shopping Location Attributes to Assess Consumer Preferences for Insect-Based Food Products. *Food Qual. Prefer.* **2017**, *55*, 45–57. [CrossRef]
59. Lensvelt, E.J.S.; Steenbekkers, L.P.A. Exploring Consumer Acceptance of Entomophagy: A Survey and Experiment in Australia and the Netherlands. *Ecol. Food Nutr.* **2014**, *53*, 543–561. [CrossRef] [PubMed]
60. Park, J.; Motoki, K.; Velasco, C.; Spence, C. Celebrity Insects: Exploring the Effect of Celebrity Endorsement on People's Willingness to Eat Insect-Based Foods. *Food Qual. Prefer.* **2022**, *97*, 104473. [CrossRef]

foods

MDPI

Article

Impact of the Russian–Ukrainian Conflict on Global Food Crops

Muh Amat Nasir [1], Agus Dwi Nugroho [2,*] and Zoltan Lakner [3]

1 Faculty of Agriculture, Universitas Gadjah Mada, Yogyakarta 55281, Indonesia
2 Doctoral School of Economic and Regional Sciences, Hungarian University of Agriculture and Life Sciences, 2100 Godollo, Hungary
3 Institute of Agricultural and Food Economics, Hungarian University of Agriculture and Life Sciences, 2100 Godollo, Hungary
* Correspondence: Nugroho.Agus.Dwi@phd.uni-mate.hu

Abstract: The Russian–Ukrainian conflict has been proven to cause significant losses of life and goods on both sides. This may have potentially impacted the agricultural sector. This study examines the impact of the conflict between Russia and Ukraine on the global food situation. We performed a descriptive analysis and literature review to answer this objective. Russia and Ukraine play essential roles in world food production and trade. However, the war has disrupted food production in Ukraine. Estimated Ukrainian wheat, soybean, and maize production in 2022–2023 fell precipitously. On the other hand, Russian production of these three food products shows positive growth during the same period. Furthermore, the global supply chain and food trade are hampered, causing an increase in the world's food prices. From March to May 2022, the average global price of wheat, soybeans, and maize increased dramatically compared to during and before the COVID-19 pandemic. Finally, this poses a danger to global food security, particularly for low-income countries that depend heavily on food imports from both countries. Therefore, all countries must be prepared for the possibility that the Sustainable Development Goals cannot be achieved.

Keywords: food security; price; production; trade

1. Introduction

The former USSR countries, such as Russia, Ukraine, and Kazakhstan, have only just begun to recover their grain production in the last two decades. However, several internal and external challenges, including institutional changes, land-use changes, climate change, and global economic trends, have significantly impacted their agricultural future. If these challenges can be overcome, the former USSR countries' grain production will increase global food security [1].

Unfortunately, the former USSR countries often experience conflict. There are ideological differences between these countries. The situation has become even more complicated since NATO expanded its influence to Central and Eastern Europe. Russia is trying to preserve its hegemony in both regions through diplomacy and confrontation [2]. Likewise, Russian and Western military technologies continue to compete with one another. As a result, there is frequently competition between the two areas for influence in several countries, especially in the EU and its neighboring countries (notably Ukraine) [3].

The conflict between Russia and Ukraine in late February 2022 exacerbated the post-COVID-19 impact on the global food situation [4]. The COVID-19 pandemic is one of the causes to blame for the price increase in 2020 and 2021 [5]. The COVID-19 pandemic put the world in a new situation where, in most cases, freedom of movement was restricted and economic, cultural, and other activities, even agricultural ones, were restricted or interrupted [6]. COVID-19 has limited the mobilization of people in terms of work, including farmers. As a result, people's incomes, welfare, and purchasing power have decreased;

food prices, food insecurity, and the number of undernourished people have increased; and the global supply chain has been disrupted [7–9]. The food supply in many countries has fallen drastically because farmers cannot sell their products, search for market information, or look for food due to the restrictions and lockdown policies [10]. Restrictions during this pandemic have also delayed agricultural activity due to labor shortages and disrupted food distribution networks [11]. On the other hand, the COVID-19 pandemic has driven a rise in global food demand [12]. This indicates an imbalance between the demand and supply of food. As a result, during the COVID-19 pandemic, food prices and food insecurity increased globally [13].

However, the global food situation was already difficult before the COVID-19 pandemic because of civil conflicts [14], trade wars [15], climate change [16], and other factors. The trade war between the US and China was the other cause of food price spikes in 2020 and 2021, whereas food trade relations increased the efficiency and stability of food prices [17]. High import tariffs, particularly those on food products, were imposed in each country during this trade conflict [15]. Consumers must bear the cost of the increased import tariffs in the form of higher food prices.

The Russian–Ukrainian conflict has been proven to cause significant losses of life and goods on both sides [4]. This may have potentially impacted the agricultural sector. Additionally, this sector in Russia and Ukraine is highly susceptible to changes. For example, harsh weather significantly affects both countries' agricultural productivity. In 2010/2011 and 2012/2013, grain yield was 30% below normal levels in Russia and 20% below average grades in Ukraine. In certain Russian districts, grain yield was more than 60% below average in 2010/2011 [18].

Based on the above, this study examines the impact of the conflict between Russia and Ukraine on the global food crops situation. Food crops are any plants that contain carbohydrates and proteins as human resources. This study only looked at three food crops: wheat, soybeans, and maize. The prices and presence of these three commodities on global commodity markets are highly volatile in the face of shocks [19]. In the next section, this study will discuss the impact of this conflict on food production, prices, trade, and security across the world.

2. Materials and Methods

This study employed annual time series data from 2020 until 2022 from various data sources, such as the United States Department of Agriculture, Food and Agriculture Organization, UN Comtrade, and Trade Map.

A descriptive or causal analysis was used to analyze the impact of the conflict between Russia and Ukraine on the global food situation. The main aspect of this analysis is that the researcher can only focus on what has happened or was happening because they have no influence over the variables. Additionally, effective research designs increase information dependability while minimizing bias [20]. Quantitative and qualitative data can be used for analysis. Descriptive studies involve quantitative data to be successful. Nevertheless, qualitative data will help in interpreting phenomena during the project investigation phase [21].

Descriptive or causal analysis can be used to understand "why" a situation has a causal effect. An excellent causal analysis can assess the effects of a situation and effectively describe population characteristics, implementation features, and the nature of the setting [22,23]. Such analysis helps the researcher in thinking systematically about many elements of a given issue, provides suggestions for more study through more analytical viewpoints and linkages, and facilitates the drawing of certain straightforward conclusions [21,24].

In addition, we conduct a literature review to provide a thorough explanation of the descriptive analysis' findings. The literature review was carried out by reading articles published in Scopus and Web of Science.

3. Results

3.1. Global Food Production

Russia is the fourth-largest wheat producer worldwide after the EU, China, and India. At the same time, Ukraine is ranked ninth, behind the United States, Canada, Australia, and Pakistan (Figure 1).

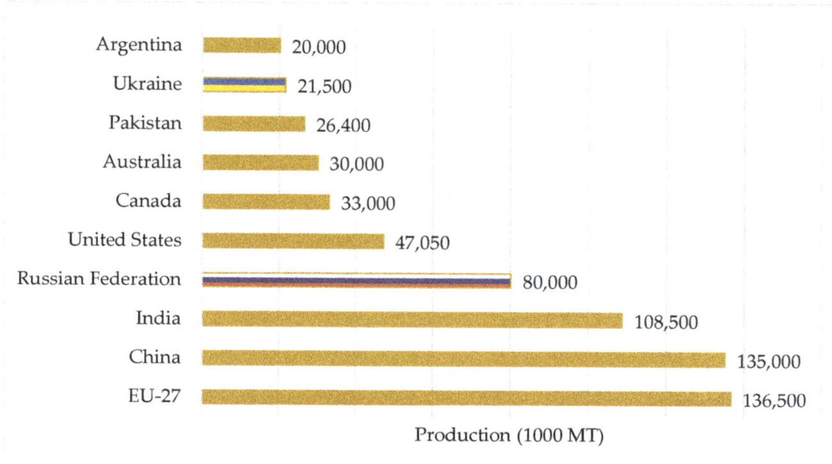

Figure 1. Estimated wheat production in 2022. Source: [25].

Figure 2 shows the top 10 soybean producers worldwide, including Russia and Ukraine. By 2022, Russia was ranked eighth globally and was projected to produce 5.3 million metric tons of soybeans. Meanwhile, Ukraine was ranked 10th and was projected to produce 2.3 million metric tons of soybeans in the same year.

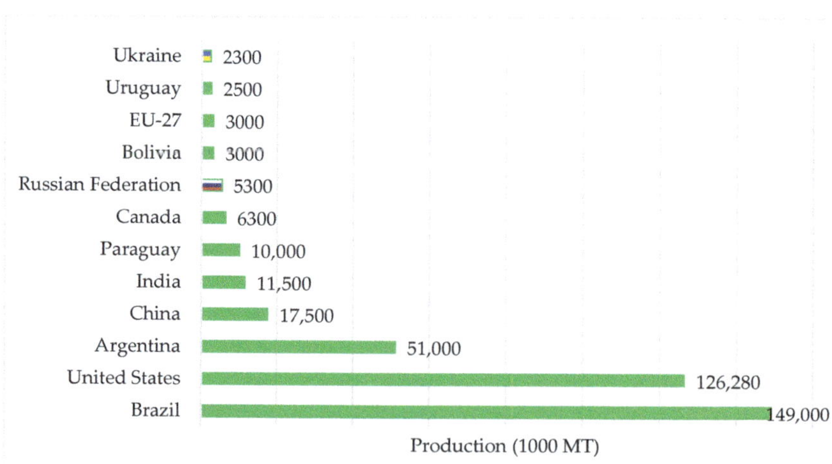

Figure 2. Estimated soybean production in 2022. Source: [25].

Russia and Ukraine are also two of the world's primary maize producers. By 2022, Russia and Ukraine will produce 15.5 million and 19.5 million metric tons of maize, respectively (Figure 3).

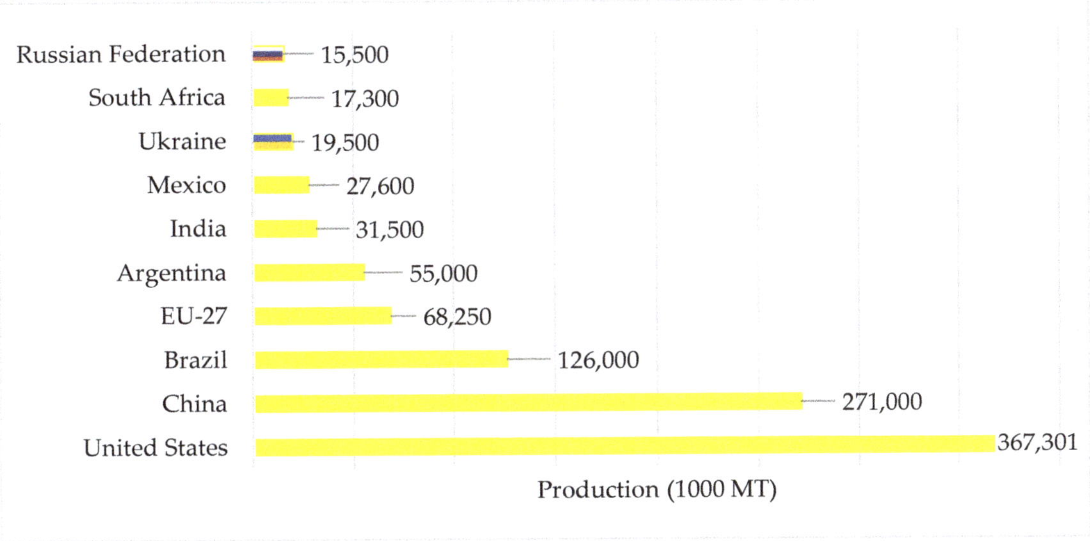

Figure 3. Estimated maize production in 2022. Source: [25].

Table 1 presents that during the conflict period, Russia's average wheat and soybean production increased (2022–2023). The reverse situation occurred in Ukraine, where the production of wheat, soybeans, and maize fell drastically.

Table 1. Average production of wheat, soybeans, and maize in Russia and Ukraine in 2017–2023.

Country	Crop	2017–2021 (000 Tons)	2021–2022 (000 Tons)	2022–2023 (000 Tons)	% Change 2022–2023
Russia	Wheat	78,194	75,158	81,500	4
	Soybean	4215	4760	5300	23
	Maize	13,598	15,225	14,500	6
Ukraine	Wheat	27,927	33,007	19,500	−26
	Soybean	4023	3800	2800	−32
	Maize	33,646	42,126	25,000	−21

Source: [25].

Ukraine is frequently referred to as the breadbasket of Europe. Additionally, it is a significant exporter of other essential commodities, including barley, rapeseed, and sunflower oil [26]. The Russian–Ukrainian conflict that began in March 2022 was a period of growing winter food crops (planted in late 2021 until early 2022) and sowing seeds for spring food crops (Figure 4).

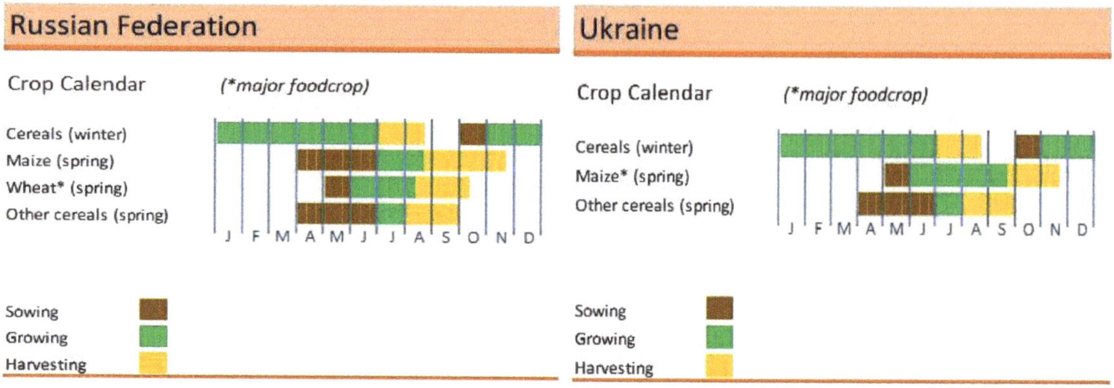

Figure 4. Russian and Ukrainian crop calendars. Source: [27]. * wheat, soybean, maize, and barley.

3.2. Global Food Price

Global food prices showed an increasing trend between May 2017 and May 2022 (Figure 5). From 2017 to 2019, global food prices had a history of stability. However, food prices began to rise in 2020 and 2021. Wheat and maize prices increased each month dramatically by 2.17% and 2.59%, respectively. During the same period, the price of soybeans increased by 1.73% per month [28].

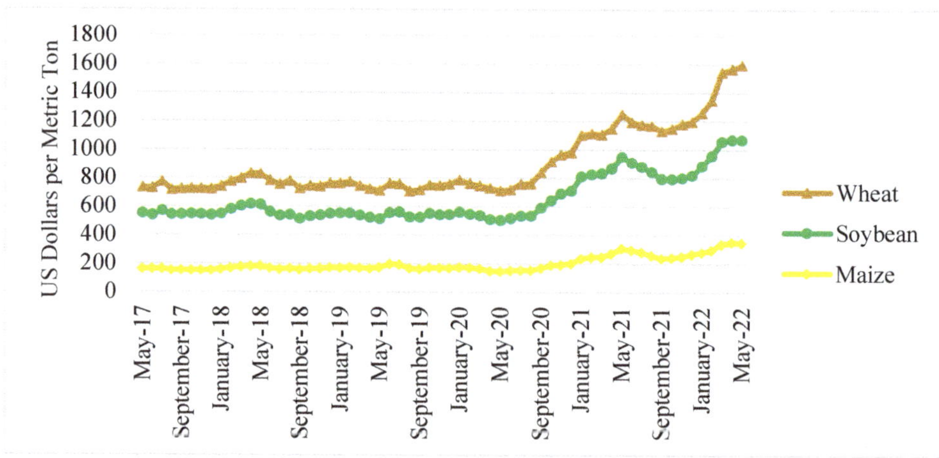

Figure 5. Global food prices from May 2017 to May 2022. Source: [28].

Global food prices increased drastically in 2022, especially in March, one month after the Russian attack on Ukraine. World soybean prices rose by 8.91% in March and rose again by around 0.03% and 0.46% in April and May, respectively. The price of maize increased higher than that of soybean, rising by 14.66% in March 2022, 3.77% in April 2022, and 0.95% in May 2022. Wheat saw the greatest price increase among all food products. This commodity's price rose to 24.53% in March and then increased again in April (1.85%) and May (5.45%). According to the FAO [29], the global food and feed prices will increase by 8% to 22% beyond their current high baseline levels if this conflict is not resolved early.

The prices of wheat, soybeans, and maize were then compared across time (Table 2). During the Russian–Ukrainian crisis, the average (mean) price of the three commodities has been at its highest. Compared to normal conditions, the average price has doubled. Meanwhile, the COVID-19 period had the most price fluctuations (standard deviation). However, given that the battle between Russia and Ukraine has been going on for a short while, things could change. The probability that price fluctuations will worsen in the future is not negligible (Table 3). According to our forecast, the price of corn will skyrocket, followed by soybeans and wheat.

Table 2. Mean and standard deviation of world wheat, soybean, and maize prices (USD per metric ton).

Period		Wheat	Soybean	Maize
January 2017–December 2019	Mean	200.04	384.27	163.44
(normal)	Std. Deviation	15.08	21.99	10.85
January 2020–February 2022	Mean	281.78	505.54	218.05
(COVID-19)	Std. Deviation	58.16	100.81	52.77
March–May 2022	Mean	501.29	721.83	342.85
(Russian–Ukrainian conflict)	Std. Deviation	15.26	1.60	5.35

Source: [28].

Table 3. Forecasting of world wheat, soybean, and maize prices (USD per metric ton).

Month	Wheat	Soybean	Maize
June 2022	347.8026	728.778	526.742
July 2022	350.3229	733.4661	564.842
August 2022	352.8432	738.1541	602.942
September 2022	355.3635	742.8421	641.042
October 2022	357.8838	747.5302	679.142
November 2022	360.4041	752.2182	717.242

Source: [28].

3.3. Global Food Trade and Security

The Russian Federation and Ukraine are significant food suppliers in the global market. The peak of wheat exports from both countries occurred in July and August. The Russian Federation was the top global wheat exporter, shipping 32.9 million tons of wheat and meslin (in product weight), or 18% of international shipments (Figure 6). With 20 million tons of wheat and meslin exported in 2021 and a 10% market share worldwide, Ukraine was the sixth-largest wheat exporter [29]. Wheat exports from both countries also continue to increase yearly.

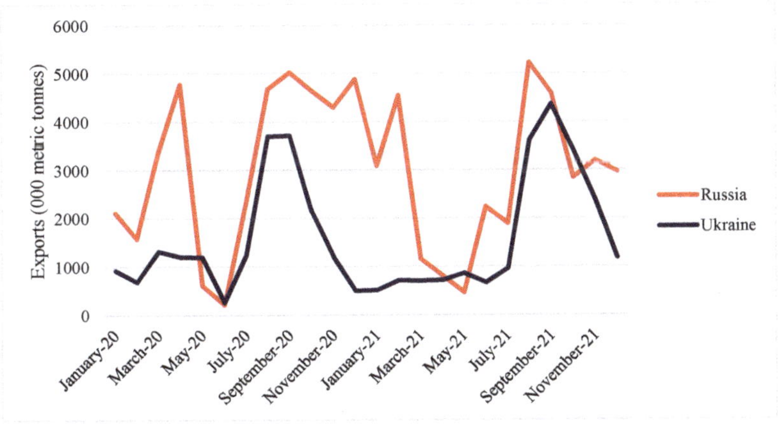

Figure 6. Export quantity of Russian and Ukrainian wheat in 2020 and 2021. Source: [30].

The final destinations of Russian wheat exports are spread across several countries in Europe, Asia, and Africa (Figure 7). The primary destinations for Russian wheat exports are Egypt, Turkey, and Bangladesh. These three countries are also Ukrainian export destinations. Russia and Ukraine provide wheat to many Least Developed Countries (LDCs) and Low-Income Food-Deficit Countries (LIFDCs). Overall, more than 30 net wheat importers in North Africa and Western and Central Asia depend on both countries for over 30% of their wheat import needs [31]. Most interestingly, Russia and Ukraine are among Yemen's sources of wheat. Yemen is in danger of famine because of a conflict, a drought, and a reliance on imported wheat [26,32]. In this sense, Yemen's food security will suffer if the supply of wheat from Russia or Ukraine is disrupted.

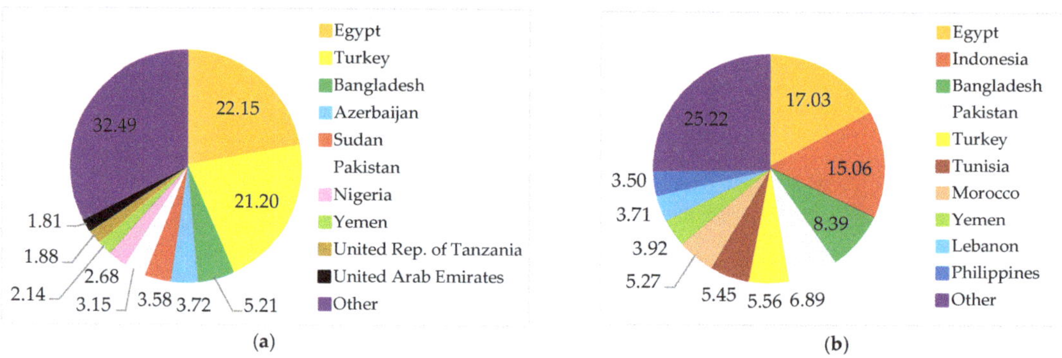

Figure 7. Destinations for Russian (**a**) and Ukrainian (**b**) wheat exports. Source: [30].

Next, we examine the proportion of wheat exported by Russia and Ukraine to the main importers. Figure 8 shows that Russia and Ukraine are primary exporters to Bangladesh, Turkey, and Egypt. About 90% of Egypt's wheat imports come from Russia and Ukraine. This confirms that the war between Russia and Ukraine will impact the importer countries' food security. This is because Russia and Ukraine cannot meet the importers' wheat needs at this time.

Our next effort was to consider the role of Russia and Ukraine as soybean exporters. Russia and Ukraine are exporters of soybeans to the world market, although in small quantities (Figure 9). Both countries' export trends indicated a slight rise by a certain amount. The peak of Ukrainian soybean exports occurred in October, while Russian soybean exports fluctuated dramatically.

Russian soybean exports mostly go to China, Belarus, Turkey, and other countries. Likewise, exports of Ukrainian soybean to these countries also showed a significant contribution (Figure 10). However, Russia and Ukraine play a far smaller role as soybean exporters than countries such as Brazil, the US, and Argentina.

The role of Ukraine and Russia was seen to be significant in fulfilling Belarusian import demands for soybeans (Figure 11). More soybeans are imported into Belarus from Ukraine than from Russia. On the other hand, both countries' contributions to fulfilling China's and Turkey's demands were less significant. Ukraine met the needs of 20.21% and 0.06%, while Russia met 2.17% and 0.69%, respectively.

Ukraine and Russia are also significant exporters of maize. The trend of maize exports from Ukraine indicates a rise in 2020 and 2021. Meanwhile, Russian maize exports mainly remained stable during that period. Ukrainian maize exports peaked at the beginning and end of the year. Meanwhile, Russian maize exports peaked in February and April (Figure 12).

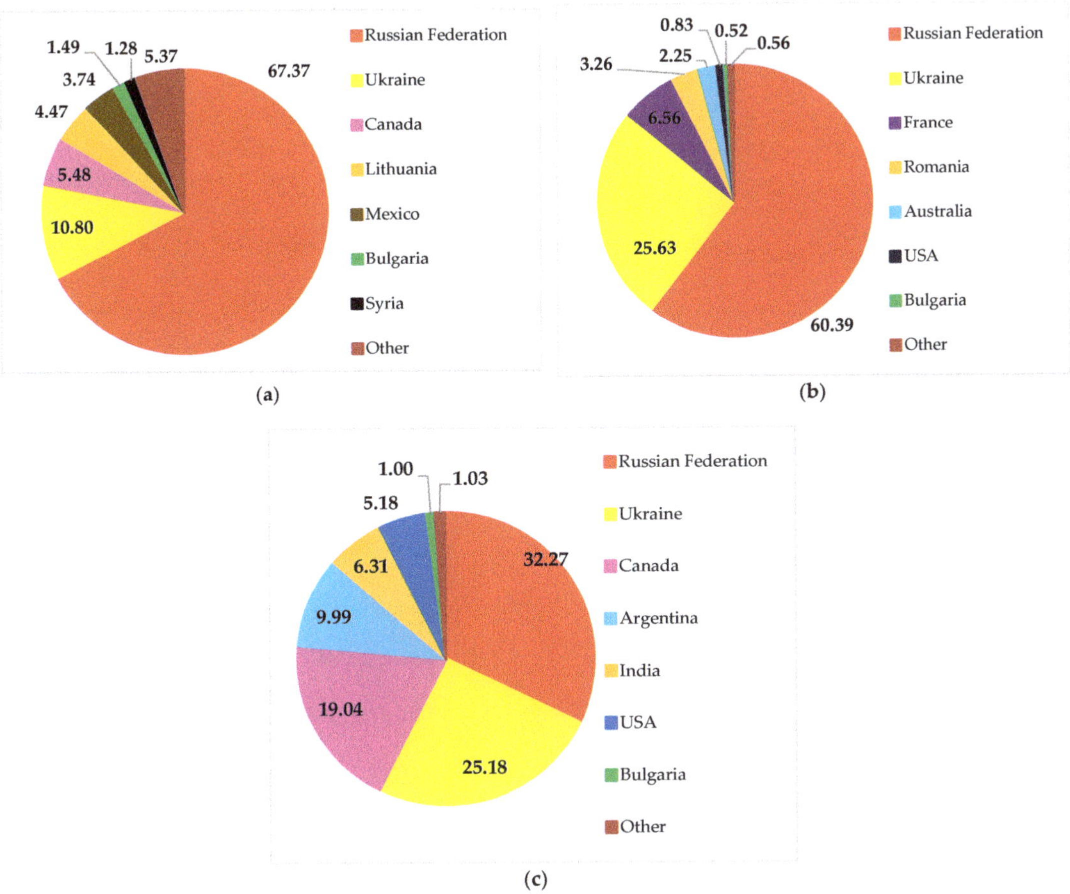

Figure 8. Turkey's (**a**), Egypt's (**b**), and Bangladesh's (**c**) wheat import structure. Source: [30].

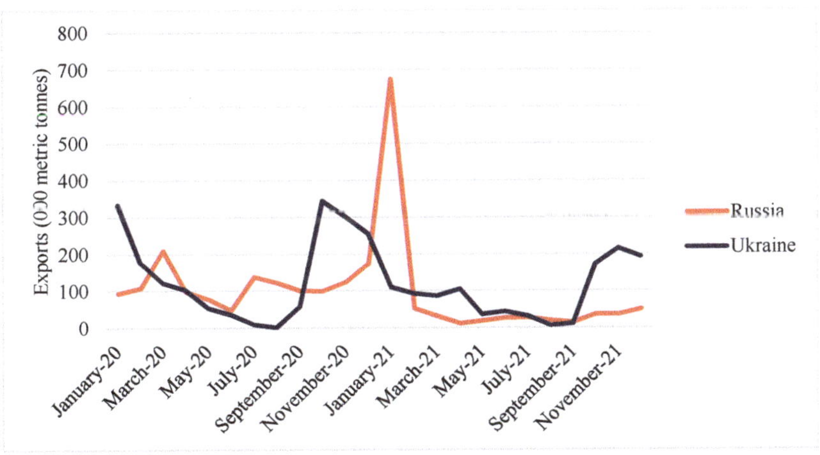

Figure 9. Export quantity of Russian and Ukrainian soybean in 2020 and 2021. Source: [30].

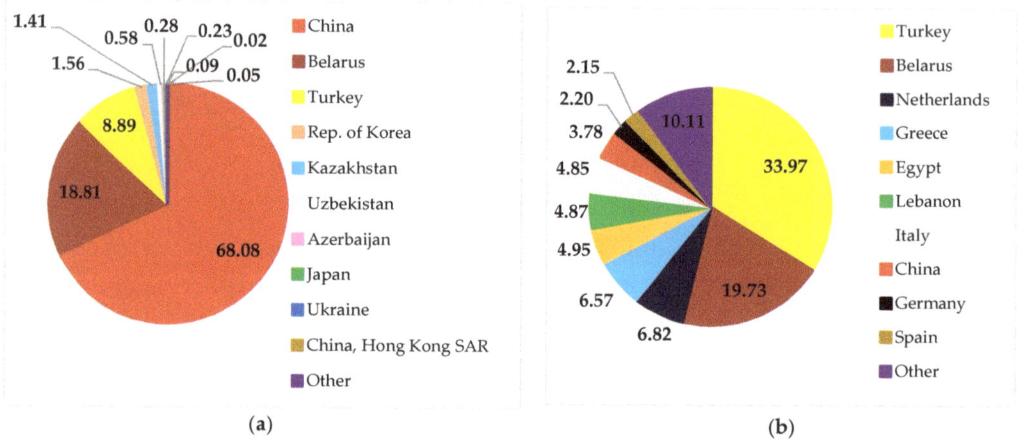

Figure 10. Destinations for Russian (**a**) and Ukrainian (**b**) soybean exports. Source: [30].

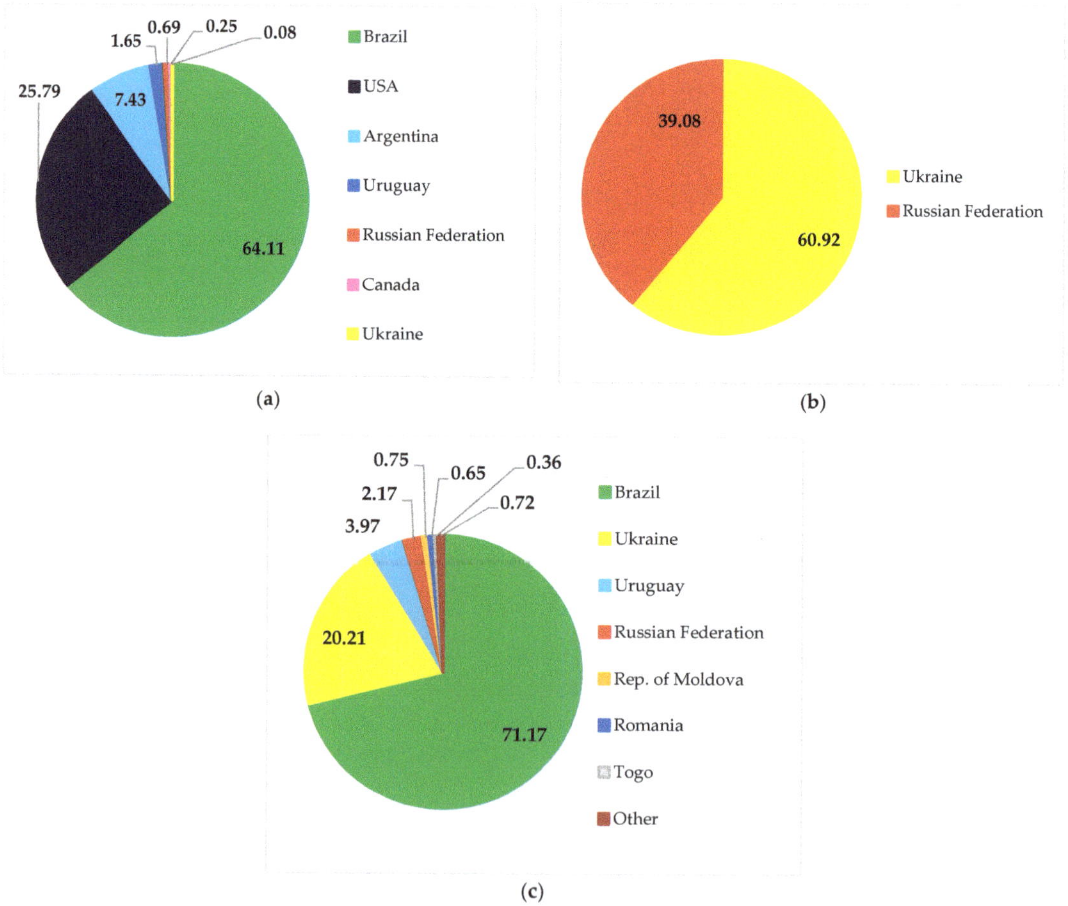

Figure 11. China's (**a**), Belarus' (**b**), and Turkey's (**c**) soybean import structure. Source: [30].

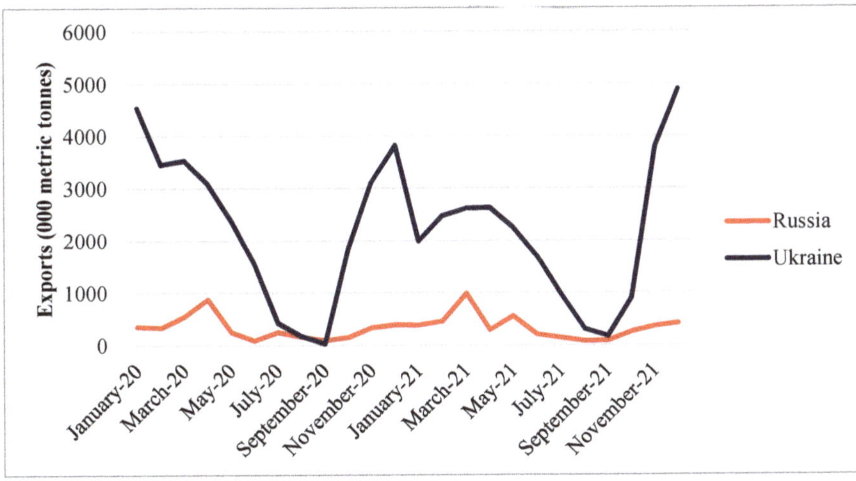

Figure 12. Export quantity of Russian and Ukrainian maize in 2020 and 2021. Source: [30].

The destination countries for maize exports from Ukraine are countries with upper-middle economies, such as China, the Netherlands, Egypt, Spain, and others. Likewise, Russia also sells maize to countries such as Turkey, Vietnam, the Republic of Korea, China, and others (Figure 13).

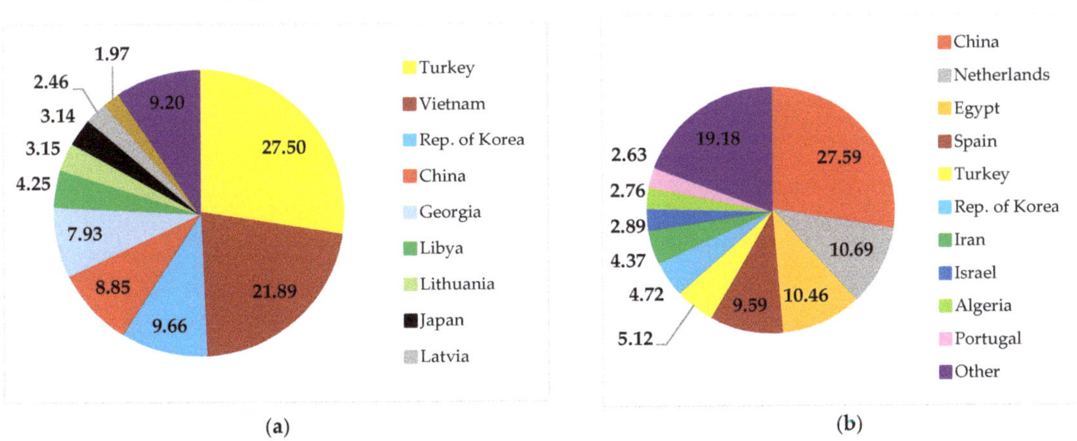

Figure 13. Destinations for Russian (**a**) and Ukrainian (**b**) maize exports. Source: [30].

We selected China, Turkey, and Korea as samples to highlight the vital role of Ukraine and Russia in the world maize trade (Figure 14). Ukraine supplies 55.55%, 30.73%, and 12.39% of maize imports to China, Turkey, and the Republic of Korea, respectively. Meanwhile, Russia is the leading supplier of Turkey's import needs (32.43%). Russia's role was smaller in meeting the import needs of the Republic of Korea (2.07%) and China (1.22%). The conflict between Ukraine and Russia has disrupted the distribution of maize to China, Turkey, and the Republic of Korea and has the potential to cause food shortages in these countries.

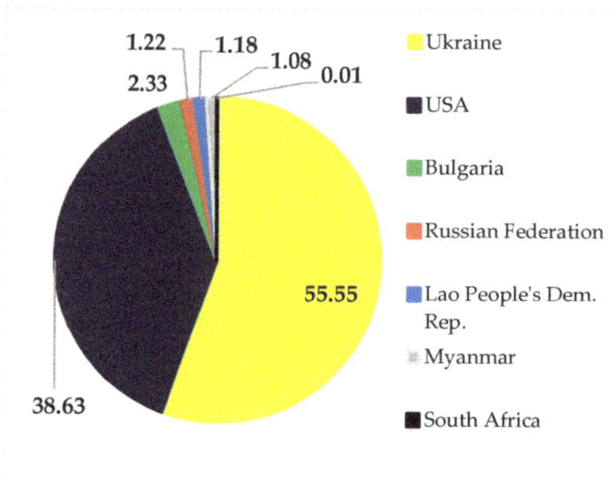

(**a**)

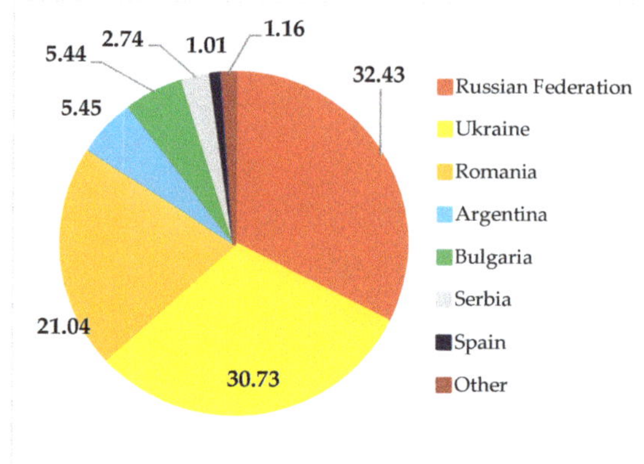

(**b**)

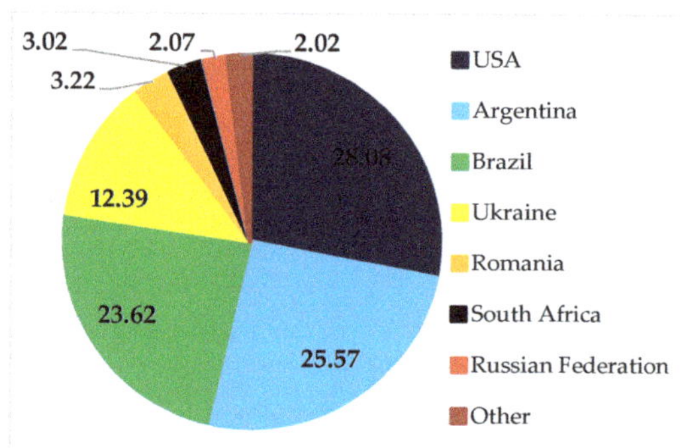

(**c**)

Figure 14. China's (**a**), Turkey's (**b**), and Korea's (**c**) maize import structure. Source: [30].

The war has generally decreased Ukrainian exports of wheat and maize (Table 4). Ukrainian exports of wheat and maize have decreased by roughly 90% during the Russian–Ukrainian conflict, as compared to pre-Russian–Ukrainian conflict values.

Table 4. Ukrainian wheat and maize exports value to the world (000 USD).

Month	Wheat		Maize	
	Mean	Std. Deviation	Mean	Std. Deviation
June 2021	167,382		669,849	
July 2021	92,477		282,569	
August 2021	348,050		193,356	
September 2021	710,911		104,548	
October 2021	755,401	285,079	84,788	384,067
November 2021	632,897		266,693	
December 2021	740,857		848,268	
January 2022	101,095		1,216,216	
February 2022	87,818		874,333	
March 2022	50,587		667,186	
April 2022	14,790	16,822	171,940	266,819
May 2022	15,016		50,220	

Source: [33].

4. Discussion

The Russian Federation and Ukraine are among the world's most important food producers. The production of food (wheat, maize, and barley) in Russia and Ukraine could rise by up to 64% (267 million tons) by 2030 [34]. Both countries have very high food crop productivity compared to other countries [35]. This is due to fertile agricultural land, sometimes known as "black soil", throughout Russia and Ukraine [36]. The significant role of Russia and Ukraine is further demonstrated by the fact that both countries' fluctuating food production is a major source of uncertainty for global food markets [31].

The roles of both countries are currently disrupted due to the war. The history of human civilization demonstrates that conflict does result in significant losses in terms of the economy, society, and human life. The conflict will significantly reduce Ukrainians' purchasing power and increase food insecurity and malnutrition. This is because people cannot engage in regular economic activity, resulting in a decreased income. The war also caused an annual average loss of 17.5% of a country's GDP per capita [37].

According to St. Augustine's Just War Theory, modern warfare will significantly increase the damage to human societies, their members, and the environment [38]. The conflict disrupts the production process for Ukraine because it prevents its farmers from attending to their fields, and harvesting and marketing their crops. Lang and McKee [26] reported that between 20% and 30% (around 1.9 million hectares) of areas sown for winter crops in Ukraine would remain unharvested during the 2022/23 season, and the next season's crops cannot be planted. The places where food production and conflict were both concentrated—such as Donetsk, Kherson, and Khrakiv—saw the most significant yield losses. Other regions of Ukraine may also experience disruptions because farmers cannot find fertilizers or control pests and diseases, and are experiencing labor shortages and a lack of storage infrastructure.

Meanwhile, Russian farmers can continue their regular agricultural activities. We can see in Table 1 that Russian food production has remained stable and even increased. The agricultural sector as a whole is still expanding favorably. However, it will be challenging to market their products owing to international economic sanctions [39]. Sanctions worsen the situation rather than improve it, with substantial adverse effects on the general population's wellbeing, including a sixfold increase in the price of food and other necessities [40]. The sanctions not only hurt Russia, but also disrupt the countries that impose them. Mardones [39] created a simulation of how sanctions will affect Russia. Economic sanctions make difficult it for Russia to export food. However, this also hurts the countries that apply sanctions, such as Latvia and Lithuania, deteriorating the condition of their food industry. Varacca and Sckokai [41] claimed that imports of food into the EU are extremely sensitive to changes in the relationship between the exporting countries and the EU. The EU will be negatively impacted if food prices in exporting countries rise. The sanctions also deal with foreign investment ceasing in Russia. At the same time, Russian agricultural investments are essential for stabilizing domestic food production and global prices [42].

According to some research, war also has increased food prices. For example, food prices have risen in Afghanistan due to the war. Since Afghanistan is culturally not the world's primary food supplier, the impact of the price rise would only be seen domestically [43]. This could be different if the war involved the world's food-producing countries, such as Russia and Ukraine. Food availability will decrease during a battle while demand stays the same or even increases. As a result, a gap between supply demand and food prices will grow [44].

According to Ali and Lin [45], the war increased food costs, which were transmitted to the consumer at a high price. In the food industry, the cost-efficiency effect is more than offset by the market power effect, resulting in lower food prices [46]. This has also occurred in the conflict between Russia and Ukraine, where there is an additional cost of food distribution due to increasing insurance premiums for ships planning to port in the Black Sea region [29]. This insurance is required to protect food distribution from war damage.

One month after the Russian–Ukrainian conflict began, in March 2022, world oil prices also rose by around 20.16%. Ultimately, this raises food prices [47]. This is reinforced by the fact that, under normal conditions or in the absence of war, fossil fuel prices have remained relatively stable for decades. This is the reason for the world's low food prices [48].

Meanwhile, the conflict will undoubtedly affect the Russian and Ukrainian exports of wheat. Our opinion is based on the evidence from the Russian and Ukrainian restrictions on wheat exports in 2007/2008 [49]. As a result, the domestic markets of Russia and Ukraine are becoming less integrated with the global wheat trade. This condition causes domestic market volatility at the micro-level, increases wheat production costs, and decreases farmer incomes. Additionally, investors will scale down and postpone planned investments in Russian and Ukrainian wheat production and infrastructure [18,50]. The ongoing conflict will also undoubtedly impair the Ukrainian and Russian role as maize suppliers to several countries, especially Turkey and China. The FAO [29] stated that this would be difficult because the potential for alternative exporters is limited to entirely replacing shipments of Ukrainian and Russian maize products. For example, Argentinian exports during the current season will also likely remain limited by government efforts to control domestic inflation, while Australia has maxed its logistical shipping capacity.

The war has already resulted in the closure of Odessa, the principal Ukrainian port, and the stopping of agricultural export activities. In addition, other modes of transportation cannot make up for the Ukrainian loss of domestic marine shipping capacity, which typically handles around 90% of the countries' commodity exports. There are also worries that the violence may harm storage and processing facilities, seaports, and inland transportation infrastructure in Ukraine [26].

Ukraine's decrease in agricultural production and export has significantly affected the scarcity of crops and grain in their import countries, particularly in LDCs and LIFDCs [4]. Moreover, these import countries have high logistical costs, increased food loss and waste, and an uncertain quality and supply of food due to inefficient infrastructure development [51]. The FAO [29] has indicated that food insecurity will increase worldwide if the Russian–Ukrainian conflict is not resolved quickly.

Wegren et al. [52] stated that the disruption of agricultural production in Eastern Europe will cause the food security of domestic and importing countries to be threatened. For example, a country such as Eritrea depends entirely on both countries to provide for its wheat needs, with Eritrea receiving 53% of its grain from Russia and 47% from Ukraine. If Ukraine cannot export wheat due to the conflict, other countries will have to meet 47% of Eritrea's wheat needs. The problem will be more complicated because the world food price is currently rising drastically. Meanwhile, Eritrea is included in the LDCs, meaning its proportion of food expenditure will increase, or its population will no longer be able to buy wheat.

This condition can be exacerbated for countries that depend on imports from Russia and Ukraine, as well as areas in conflict, such as Afghanistan. Afghanistan's people are food-insecure due to the war from the early 2000s. This was brought on by a decline in food production, job losses, reduced income, and a rise in food prices [43]. Eventually, Eritrea, Afghanistan, and other countries that depend on wheat imports from Russia and Ukraine will experience food insecurity.

Even the FAO's simulations suggest that under such a scenario, the global number of undernourished people could increase by 8 to 13 million in 2022/2023, with the most pronounced increases taking place in Asia-Pacific, followed by sub-Saharan Africa, the Near East, and North Africa [29]. This means that the world must be ready to deal with health issues brought on by food shortages in the future, such as stunting in children, poor quality of life, reduced capability for earning an income, illness, and mortality [48]. As a result, this will hinder the achievement of the SDGs and an adequate food system.

5. Conclusions

The conflict between Russia and Ukraine has harmed both of these countries as well as other countries. This conflict has caused the food production capacity of Ukraine to decline. Agricultural activity cannot be conducted as usual by Ukrainians. Infrastructure for the processing and distribution of food has been destroyed and rendered inoperable. Additionally, the main ports in Ukraine are blockaded. This differs from the Russian agricultural sector, which, as a whole, is still expanding favorably. Therefore, the conflict has not negatively impacted Russia, at least in terms of fundamental agricultural production. In the 2022–2023 season, Russia is also projected to export more wheat than ever before. The most pressing problem for Russia is that several countries have imposed economic sanctions, causing the country to experience barriers to exporting food products.

One month after the conflict began, in March 2022, there was a dramatic increase in the price of food throughout the world. The maize price increased by 14.66%, the soybeans price increased by 8.91%, and the wheat price increased by 24.53%. The world was concerned about this situation, mainly Russian and Ukrainian food importers, which include LDCs and LIFDCs.

This situation will threaten the achievement of the Sustainable Development Goals (SDGs), especially zero hunger. Even the FAO has estimated that 8 to 13 million more people might become undernourished globally in 2022–2023 [29], a condition that will start a domino effect that worsens the quality of health and life, increases poverty and inequality, slows economic growth, and contradicts the SDGs.

The only thing we can recommend is that all countries, notably Russia and Ukraine, use diplomacy to stop this conflict. Both Russia and Ukraine, as well as other countries, stand to gain nothing from this confrontation. Economic sanctions have also made the situation worse. Russia must stop the war and respect Ukraine's desires. Similarly, Ukraine should prioritize strengthening relations with Russia and attempting to take a stance while avoiding the influence of other countries. We also encourage the United Nations to initiate a discussion attended by only the leaders of Russia and Ukraine to negotiate in a neutral place. Countries that support Russia and Ukraine must also intervene to bring about peace, rather than escalating the confrontation between the two. In the short term, we urge Russia to open the port in Ukraine to supply wheat, soybeans, and maize to other countries. Russia and Ukraine may involve the United Nations Peacekeeping Forces in these activities to maintain logistic transportation safety and to assuage both countries' suspicions.

The issue of Ukrainian refugees and isolated populations must also be prioritized by UNHCR in order for them to have access to adequate food. Then, the FAO and all countries must collaborate to strengthen the food supply chain when this war is over to make it more effective and efficient. Our study's main limitation is the scarcity of data due to the recent Russia–Ukraine conflict. As a result, we encourage future research on the impact of this conflict, including more comprehensive data.

Author Contributions: Conceptualization, M.A.N., A.D.N. and Z.L.; methodology, A.D.N.; software, A.D.N.; validation, A.D.N. and Z.L.; formal analysis, A.D.N.; investigation, A.D.N.; resources, A.D.N.; data curation, A.D.N.; writing—original draft preparation, M.A.N. and A.D.N.; writing—review and editing, M.A.N. and A.D.N.; visualization, A.D.N.; supervision, Z.L. All authors have read and agreed to the published version of the manuscript.

Funding: This research received no external funding.

Institutional Review Board Statement: Not applicable.

Informed Consent Statement: Not applicable.

Data Availability Statement: The data presented in this study are available on request from the corresponding author.

Acknowledgments: We would like to thank the reviewers who gave us suggestions on how to develop this article.

Conflicts of Interest: The authors declare no conflict of interest.

References

1. Lioubimtseva, E.; Henebry, G.M. Grain Production Trends in Russia, Ukraine and Kazakhstan: New Opportunities in an Increasingly Unstable World? *Front. Earth Sci.* **2012**, *6*, 157–166. [CrossRef]
2. Goldgeier, J.; Itzkowitz Shifrinson, J.R. *Evaluating NATO Enlargement: Scholarly Debates, Policy Implications, and Roads Not Taken*; Palgrave Macmillan: London, UK, 2020; Volume 57, ISBN 0123456789.
3. Kupriyanov, A.V. Cold War as a Special Type of Conflict: A Strategic Sketch. *Russ. Glob. Aff.* **2022**, *20*, 78–92. [CrossRef]
4. Shams Esfandabadi, Z.; Ranjbari, M.; Scagnelli, S.D. The Imbalance of Food and Biofuel Markets amid Ukraine-Russia Crisis: A Systems Thinking Perspective. *Biofuel Res. J.* **2022**, *9*, 1640–1647. [CrossRef]
5. Dasgupta, S.; Robinson, E.J.Z. Impact of COVID-19 on Food Insecurity Using Multiple Waves of High Frequency Household Surveys. *Sci. Rep.* **2022**, *12*, 1865. [CrossRef] [PubMed]
6. Hambardzumyan, G.; Gevorgyan, S. The Impact of COVID-19 on the Small and Medium Dairy Farms and Comparative Analysis of Customers' Behavior in Armenia. *Future Foods* **2022**, *5*, 100110. [CrossRef]
7. Khan, S.A.R.; Razzaq, A.; Yu, Z.; Shah, A.; Sharif, A.; Janjua, L. Disruption in Food Supply Chain and Undernourishment Challenges: An Empirical Study in the Context of Asian Countries. *Socio-Econ. Plan. Sci.* **2022**, *82*, 101033. [CrossRef]
8. Onyeaka, H.; Ukwuru, M.; Anumudu, C.; Anyogu, A. Food Fraud in Insecure Times: Challenges and Opportunities for Reducing Food Fraud in Africa. *Trends Food Sci. Technol.* **2022**, *125*, 26–32. [CrossRef]
9. Nugroho, A.D. Does Covid-19 Have Effects on the Indonesian Horticultural Subsector? *Bulg. J. Agric. Sci.* **2021**, *27*, 865–874.
10. Adhikari, J.; Timsina, J.; Khadka, S.R.; Ghale, Y.; Ojha, H. COVID-19 Impacts on Agriculture and Food Systems in Nepal: Implications for SDGs. *Agric. Syst.* **2021**, *186*, 102990. [CrossRef]
11. Rajpoot, K.; Singla, S.; Singh, A.; Shekhar, S. Impact of COVID-19 Lockdown on Prices of Potato and Onion in Metropolitan Cities of India. *J. Agribus. Dev. Emerg. Econ.* **2022**, *12*, 386–399. [CrossRef]
12. Soares, J.B.; da Costa, M.R.; Monteiro-Neto, C.; Loto, L.; de Abreu, M.D.; Tubino, R.D.A. Impacts of COVID-19 on the Value Chain of a Small-Scale Fishery System in a Tropical Metropolitan City. *Mar. Policy* **2022**, *140*, 105068. [CrossRef]
13. Zhang, C.; Yang, Y.; Feng, Z.; Xiao, C.; Lang, T.; Du, W.; Liu, Y. Risk of Global External Cereals Supply under the Background of the Covid-19 Pandemic: Based on the Perspective of Trade Network. *Foods* **2021**, *10*, 1168. [CrossRef] [PubMed]
14. Biggeri, M.; Cuesta, J.; Ferrone, L.; Hamza Abbas, M.; Khurshid, A. Children in the Context of War: Deprivation among Internally Displaced, Returnee, Host and Stayee Children in East Mosul. *J. Dev. Stud.* **2022**, *58*, 1032–1052. [CrossRef]
15. Ge, J.; Polhill, J.G.; Macdiarmid, J.I.; Fitton, N.; Smith, P.; Clark, H.; Dawson, T.; Aphale, M. Food and Nutrition Security under Global Trade: A Relation-Driven Agent-Based Global. *R. Soc. Open Sci.* **2021**, *8*, 201587. [CrossRef] [PubMed]
16. Luo, Q.; Hoogenboom, G.; Yang, H. Uncertainties in Assessing Climate Change Impacts and Adaptation Options with Wheat Crop Models. *Theor. Appl. Climatol.* **2022**, *149*, 805–816. [CrossRef]
17. Wang, J.; Dai, C. Evolution of Global Food Trade Patterns and Its Implications for Food Security Based on Complex Network Analysis. *Foods* **2021**, *10*, 2657. [CrossRef]
18. Götz, L.; Djuric, I.; Nivievskyi, O. Regional Price Effects of Extreme Weather Events and Wheat Export Controls in Russia and Ukraine. *J. Agric. Econ.* **2016**, *67*, 741–763. [CrossRef]
19. Greb, F.; Prakash, A. *Assessing Volatility Patterns in Food Crops*; FAO: Rome, Italy, 2017.
20. Kothari, C.R. *Research Methodology Methods and Techniques*, 2nd ed.; New Age International Publishers: New Delhi, India, 2004; Volume 1999, ISBN 9788122424881.
21. Sekaran, U.; Bougie, R. *Research Methods For Business: A Skill Building Approach*, 7th ed.; John Wiley and Sons: New York, NY, USA, 2016; Volume 26.
22. Pandey, P.; Pandey, M.M. *Research Methodology: Tools and Techniques*; Bridge Center: Buzau, Romania, 2015; ISBN 9786069350270.
23. Loeb, S.; Dynarski, S.; McFarland, D.; Morris, P.; Reardon, S.; Reber, S. *Descriptive Analysis in Education: A Guide for Researchers*; Department of Education, Institute of Education Sciences, National Center for Education Evaluation and Regional: Washington, DC, USA, 2017.
24. Kielmann, K.; Cataldo, F.; Seeley, J. *Introduction to Qualitative Research Methodology: A Training Manual*; Department for International Development (DfID): London, UK, 2012.
25. United States Department of Agriculture. USDA Database. Available online: https://www.usda.gov/topics/data (accessed on 30 June 2022).
26. Lang, T.; McKee, M. The Reinvasion of Ukraine Threatens Global Food Supplies. *BMJ* **2022**, *376*, 10–11. [CrossRef]
27. Food and Agriculture Organization of the United Nations. Fao Crop Calendar. 2022. Available online: https://data.apps.fao.org/catalog/dataset/crop-calendar-by-country-crop-and-activity (accessed on 1 July 2022).
28. World Bank. World Bank Open Data. Available online: https://data.worldbank.org/ (accessed on 30 June 2022).
29. FAO. *The Importance of Ukraine and the Russian Federation for Global Agricultural Markets and the Risks Associated with the Current Conflict*; FAO: Rome, Italy, 2022.
30. United Nations Statistics Division. UN Comtrade Database. 2022. Available online: https://comtrade.un.org/ (accessed on 1 July 2022).
31. Araujo-Enciso, S.R.; Fellmann, T. Yield Variability and Harvest Failures in Russia, Ukraine and Kazakhstan and Their Possible Impact on Food Security in the Middle East and North Africa. *J. Agric. Econ.* **2020**, *71*, 493–516. [CrossRef]

32. The Economist Group. Global Food Security Index (GFSI). Available online: https://impact.economist.com/sustainability/project/food-security-index/Downloads (accessed on 2 July 2022).
33. International Trade Center. Trade Statistics for International Business. 2022. Available online: https://www.trademap.org/Index.aspx (accessed on 2 July 2022).
34. Deppermann, A.; Balkovič, J.; Bundle, S.C.; Di Fulvio, F.; Havlik, P.; Leclère, D.; Lesiv, M.; Prishchepov, A.V.; Schepaschenko, D. Increasing Crop Production in Russia and Ukraine-Regional and Global Impacts from Intensification and Recultivation. *Environ. Res. Lett.* **2018**, *13*, 025008. [CrossRef]
35. Svanidze, M.; Đurić, I. Global Wheat Market Dynamics: What Is the Role of the Eu and the Black Sea Wheat Exporters? *Agriculture* **2021**, *11*, 799. [CrossRef]
36. Zhao, Z.; Fang, X.; Ye, Y.; Zhang, C.; Zhang, D. Reconstruction of Cropland Area in the European Part of Tsarist Russia from 1696 to 1914 Based on Historical Documents. *J. Geogr. Sci.* **2020**, *30*, 1307–1324. [CrossRef]
37. Costalli, S.; Moretti, L.; Pischedda, C. The Economic Costs of Civil War: Synthetic Counterfactual Evidence and the Effects of Ethnic Fractionalization. *J. Peace Res.* **2017**, *54*, 80–98. [CrossRef]
38. Langan, J. The Elements of St. Augustine ' s Just War Theory. *J. Relig. Ethics* **1984**, *12*, 19–38.
39. Mardones, C. Economic Effects of Isolating Russia from International Trade Due to Its 'Special Military Operation' in Ukraine. *Eur. Plan. Stud.* **2022**, 1–16. [CrossRef]
40. Sen, K.; Faisal, W. Al Public Health Challenges in the Political Economy of Conflict: The Case of Syria. *Int. J. Health Plan. Manag.* **2015**, *30*, 314–329. [CrossRef]
41. Varacca, A.; Sckokai, P. The Determinants of the EU Import Demand for Soybean and Maize: What Role for GMOS? *J. Agric. Food Ind. Organ.* **2020**, *18*, 1–16. [CrossRef]
42. Koizumi, T. Impact of Agricultural Investments on World Wheat Market under Climate Change: Effects of Agricultural Knowledge and Innovation System, and Development and Maintenance of Infrastructure. *Jpn. Agric. Res. Q.* **2019**, *53*, 109–125. [CrossRef]
43. Gohar, A.A.; Ward, F.A.; Amer, S.A. Economic Performance of Water Storage Capacity Expansion for Food Security. *J. Hydrol.* **2013**, *484*, 16–25. [CrossRef]
44. Hussein, M.; Law, C.; Fraser, I. An Analysis of Food Demand in a Fragile and Insecure Country: Somalia as a Case Study. *Food Policy* **2021**, *101*, 102092. [CrossRef]
45. Ali, H.E.; Lin, E.S. Wars, Foodcost and Countervailing Policies: A Panel Data Approach. *Food Policy* **2010**, *35*, 378–390. [CrossRef]
46. Azzam, A.; Andersson, H. Measuring Price Effects of Concentration in Mixed Oligopoly: An Application to the Swedish Beef-Slaughter Industry. *J. Ind. Compet. Trade* **2008**, *8*, 21–31. [CrossRef]
47. War, G. War and the Global Economy. *Econ. Outlook* **2002**, *27*, 9–14. [CrossRef]
48. Darnton-Hill, I.; Cogill, B. Maternal and Young Child Nutrition Adversely Affected by External Shocks Such as Increasing Global Food Prices. *J. Nutr.* **2010**, *140*, 162S–169S. [CrossRef] [PubMed]
49. Kopytko, N. Change and Transition: The Climate of Ukraine's Agri-Food Sector. *Clim. Policy* **2016**, *16*, 68–87. [CrossRef]
50. Götz, L.; Glauben, T.; Brümmer, B. Wheat Export Restrictions and Domestic Market Effects in Russia and Ukraine during the Food Crisis. *Food Policy* **2013**, *38*, 214–226. [CrossRef]
51. Perdana, T.; Chaerani, D.; Hermiatin, F.R.; Achmad, A.L.H.; Fridayana, A. Improving the Capacity of Local Food Network through Local Food Hubs' Development. *Open Agric.* **2022**, *7*, 311–322. [CrossRef]
52. Wegren, S.K.; Nikulin, A.M.; Trotsuk, I. The Fragility of Russia's Agricultural Production and Implications for Food Security. *Eurasian Geogr. Econ.* **2021**, 1–39. [CrossRef]

foods

MDPI

Article

Spatiotemporal Differences, Dynamic Evolution and Trend of the Coupled Coordination Relationship between Urbanization and Food Security in China

Shan Liu [1] **and Mengyang Hou** [2,3,*]

[1] School of Management, Shijiazhuang Tiedao University, Shijiazhuang 050043, China
[2] School of Economics, Hebei University, Baoding 071000, China
[3] Research Center of Resources Utilization and Environmental Conservation, Hebei University, Baoding 071000, China
* Correspondence: houmengyang@hbu.edu.cn

Abstract: Scientific assessment of the coupled coordination degree between urbanization and food security (CDUFS) revealed regional differences and sources. Dynamic evolution and trends are important references for achieving a coordinated interaction between high-quality urbanization and ensuring food security. Specifically, the CDUFS was measured using prefectural panel data in China from 2000 to 2019 and the coupling coordination degree model, which revealed its spatial correlation and differentiation. On this basis, in order to examine the spatiotemporal differences and evolution of the CDUFS, the Dagum–Gini coefficient and Kernel density estimation were innovatively used to analyze its regional differences and evolution distribution. The spatial Markov chain was further employed to examine the evolution trend of the CDUFS. The study found that the CDUFS showed a downward trend in fluctuation within the low coordination interval. There was a positive spatial correlation, with a more stable distribution pattern of high–high and low–low clusters. The regional differences in the CDUFS were obvious and the overall difference has expanded. The main source of regional differences among different food functional areas was inter-regional differences, followed by intra-regional differences. The regional difference between food main producing areas and food main marketing areas was the highest. The CDUFS shows a single-peak distribution; the imbalance between regions was still prominent with a left trailing phenomenon and no convergence. The CDUFS has the stability of maintaining the original state, and the probability of leapfrogging evolution is low in the short term. Finally, the geospatial effect plays an important role in the dynamic evolution of the CDUFS.

Keywords: urbanization; food security; regional differences; dynamic distribution; evolution trend; Dagum–Gini coefficient; kernel density estimation; spatial Markov chain

1. Introduction

One of the global Sustainable Development Goals (SDGs) set by the United Nations is to achieve food security and promote sustainable agriculture [1]. Ensuring food security is also an important foundation to promote urbanization. Since the reform and opening up, China has experienced an unprecedented rapid urbanization process [2]. The rate of population urbanization increased from 17.92% in 1978 to 60.60% in 2019, with an average annual increase of 3.02%. Food security, characterized by grain yield per capita, only increased from 316.61 kg/person to 470.78 kg/person, with an average annual increase of only 0.97%, significantly lagging behind urbanization. China has the ability to guarantee the supply of grain and important agricultural products, but grain production is facing many challenges such as rising costs [3,4], resource and environmental constraints [5,6], growth in rigid demand under domestic and international market changes, climate change, etc. [7]. At present, grain production of China has stopped in 2015 at the "twelfth consecutive increase", and there is still supply pressure [8] with a tight balance of grain supply and demand [9].

Urbanization is mainly manifested by the large-scale movement of the rural population and the demand of land expansion [10], which leads to the structural adjustment of inputs, resulting in a decrease of cropland and structural shortage of rural labor. Cropland use is also gradually non-agricultural [11]. Urbanization also leads to the growth of farm households' non-farm income and the consumption diversification of agro-foods, resulting in the non-farming of food production [12]. Food security means maximizing the stability of food supply and ensuring that anyone has access to the food they need for survival and health, while non-agricultural and non-farming threaten food supply per capita and further affect food security. To ensure food security, adequate factor inputs and a reasonable factor allocation structure must be met, which will place a higher demand on urbanization development. If too much emphasis is placed on ensuring food security, it may cause a mobility barrier and a spatial mismatch of the urban-rural factor structure [13,14]; the development of urbanization would be unsustainable. In short, the one-sided pursuit of urbanization expansion would be a strain to food security [15]. Conversely, if food security is emphasized too much, the factor support for urbanization may be insufficient. There may be a mutually equilibrium relationship between urbanization and food security [16,17]. With the transition to high-quality socioeconomic development, a stable and coordinated state must exist between the two systems. So, it is necessary to objectively measure the coordinated relationship between urbanization and food security in China, and deeply explore its spatial-temporal differentiation, regional differences, dynamic evolution and trends. It is expected that this study can provide reference for the sustainable development of urbanization and food security in China and other developing countries.

Existing studies have launched rich exploration of the relationship between urbanization and food security, but mainly focused on the unidirectional impact of urbanization [17]. The unbalanced development of urbanization has created real problems and practical contradictions in food security [11], but urbanization from different perspectives have different impacts on food security. Xu found that economic urbanization and population urbanization in Jiangsu province can promote food security, while land urbanization and consumption urbanization have different degrees of negative effects on food security [18]. While a study at the national level found that urbanization has a significant positive impact on food security, the impact varies across food functional areas, with urbanization in the main food marketing areas having a significant negative impact on food security [19]. In terms of the two-way interaction between urbanization and food security, Yao et al. quantitatively analyzed the relationship between the urbanization system and food security system, the relationship between the two systems is increasing [20]. Zhu et al. measured the coupling coordination between new urbanization and food security in Henan province and found that the coordination degree of the two systems in Henan province basically achieved a leap of "near disorder → moderate coordination" [21]. Hou et al. believed that the coupling coordination between urbanization and grain production in China has just entered a highly coordinated range, and there is still much room for improvement [22]. Meanwhile, the study of the relationship between the two systems involves the interaction between urbanization and cultivate land pressure, total factor productivity [23–25], coordinated development between agricultural modernization and urbanization [26,27].

In general, the existing studies provide a solid theoretical foundation for an exploration of the relationship between these two systems. However, it is still insufficient to excavate the change laws, so there is still room for exploration. First, regarding research scale, most of the literature focuses on analysis at the provincial level [20] or in a particular province, which may not accurately reflect the imbalance degree of urbanization and food security within province. Second, regarding research content, most of the literature analyzed the one-way impact of urbanization on food security but there is still a lack of systematic study on the coordination between the two. Third, regarding research depth, the studies on spatial-temporal differences mostly use GIS technology to compare and analyze the different spatial distribution of time-series variability [21,22]. The examination of regional differences and evolution trend is still insufficient.

In total, this study examines in more detail on a smaller dimension at the prefecture level [28] to overcome the homogenization error due to provincial macro data. Specifically, using the panel data of 330 prefecture-level and above cities in China from 2000–2019 as the research sample. Based on measurement of the comprehensive level of urbanization by the entropy method, the coordination degree between urbanization and food security (CDUFS) is measured by the coupled coordination model and reveals its spatial-temporal differentiation through the exploratory spatial data analysis (ESDA). From the perspective of different food functional areas, the regional differences and distribution dynamics of the CDUFS are analyzed by the Dagum–Gini coefficient and Kernel density estimation, and then the spatial Markov chain method is used to analyze its evolution trend. The study purpose was to grasp the evolution law of the interaction between urbanization and food security in China at the macro level. This is an important theoretical reference for other developing countries facing food security problems in the process of urbanization and for achieving a balance relationship between economic development and food security.

2. Materials and Methods

2.1. Variable Definition

In order to scientifically assess the coordinating relationship between urbanization and food security, a clear definition and measurement of both is needed. Specifically:

(1) Urbanization. Urbanization is a comprehensive system with multi-dimensional features such as population mobility, urban land growth and non-agricultural economy development [20]. A single-dimensional population urbanization is different to reflect the complex characteristics of urbanization [29,30]. With reference to existing studies and data acquisition at the prefecture-level, indicators are selected from population, land, and economy to assess the comprehensive level of urbanization [31,32]. Population urbanization is characterized by the urbanization rate of resident population. Land urbanization is equal to the proportion of the urban built-up area to land area. Economy urbanization is equal to the proportion of non-agricultural industries to GDP. The entropy method is used to assign weights to indicators and conduct comprehensive evaluation to avoid the subjectivity of artificial assignment [33]. Limited to space, the detailed calculation can be found in the cited literature. The relevant indicators are standardized by the polarization method before calculation [34].

(2) Food security. Food security is also a comprehensive system containing multi-dimensional indicators such as self-sufficiency rate, reserve level, and per capita food possession [35,36]. However, some indicator data are difficult to obtain at the prefecture level, so from the perspective of per capita output scale, we use food output per capita (food production/total population) to portray food security. The reason is that the fundamental strategy to ensure food security remains to increase the output capacity per capita of food to stabilize supply. In addition, the international food security standard line is also measured by per capita food possession [37].

2.2. Data Sources

The study sample in this paper is panel data of 330 prefecture-level and above cities from 2000–2019. Tibet, Hong Kong, Macao and Taiwan are not involved due to the limitation of data availability. The data involved are mainly obtained from the China Urban Statistical Yearbook, China Regional Economic Statistical Yearbook and provincial statistical yearbooks (https://data.cnki.net/Yearbook, accessed on 20 May 2022). The missing data are supplemented by prefecture—level municipal statistical yearbooks and statistical bulletins. The land use data is obtained from the European Space Agency Climate Change Initiative (CCI) global land cover product data (www.esa-landcover-cci.org, accessed on 20 May 2022) with spatial-temporal resolutions of annual scale and 300 m × 300 m [38]. Table 1 shows the brief descriptive statistics.

Table 1. Brief descriptive statistics.

Variables			Variable Desceition	Mean	Std.
CDUFS	Urbanization	Population urbanization	urbanization rate of resident population	0.2824	0.0631
		Land urbanization	urban built-up area/land area		
		Economy urbanization	non-agricultural industries/GDP		
	Food security	Food output per capita	food production/total population		

In addition, to examine the regional differences of the CDUFS, the provinces where the prefecture-level cities are located nationwide are divided into three different food functional areas (Figure 1), food main producing areas (FPAs), food main marketing areas (FMAs) and food balanced areas (FBAs), according to different agricultural production endowments and subjects [22].

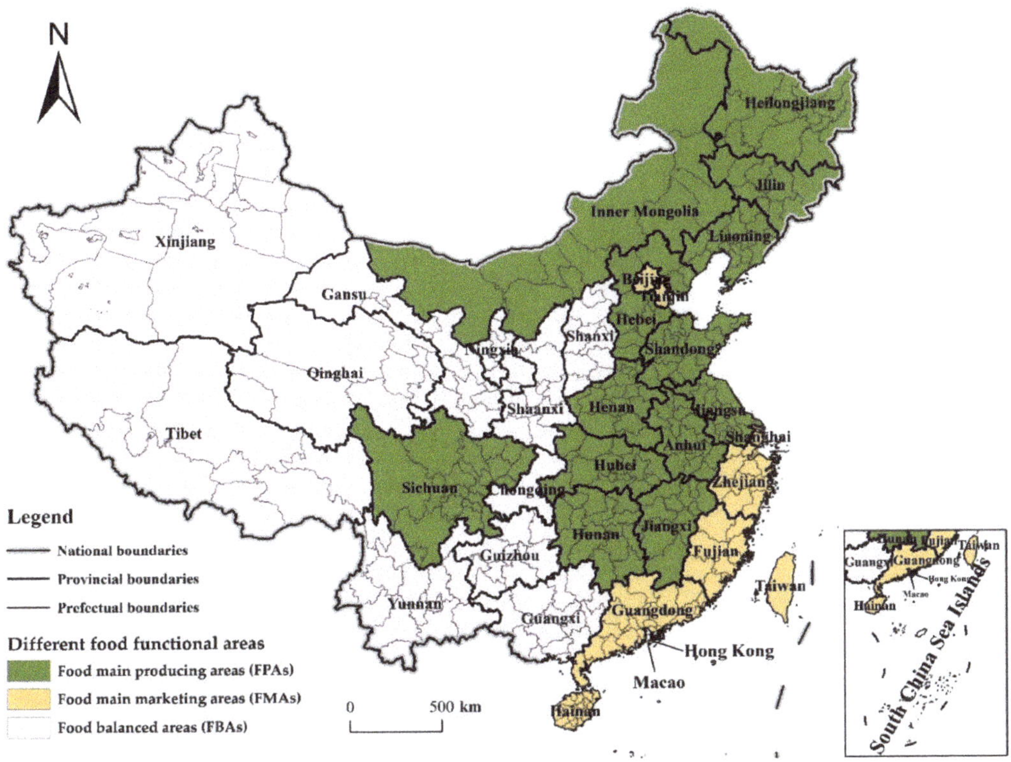

Figure 1. The distribution of different food functional areas. Note: This figure is drawn by the authors themselves based on ArcGIS.

2.3. Research Methods

2.3.1. CDUFS Measurement: Coupling Coordination Degree Model

The coordination degree between urbanization and food security (CDUFS) is measured by the coupling coordination degree model. The equation is as follows.

$$C = 2 * \left\{ \frac{U_1 \times U_2}{(U_i + U_j)^2} \right\}^{1/2} \quad (1)$$

$$\begin{cases} D = \sqrt{C \times T} \\ T = aU_1 + bU_2 \end{cases} \tag{2}$$

In Equations (1) and (2), C is the coupling degree, and $C \in [0,1]$; U_1, U_2 are urbanization and food security, respectively. D is the CDUFS, T is the comprehensive coordination index between urbanization and food security, reflecting the synergistic effect of the two, and $T \in (0, 1)$ to ensure $D \in (0, 1)$. A, b are the pending coefficients, and $a + b = 1$.

Treating urbanization and food security as equally important, set $a = 0.5$, $b = 0.5$. D is classified into four coordination types of Low ($0 \leq D < 0.3$), Moderate ($0.3 \leq D < 0.5$), High ($0.5 \leq D < 0.8$), and Extreme ($0.8 \leq D < 1$) [39].

2.3.2. Spatiotemporal Characteristics: Exploratory Spatial Data Analysis (ESDA)

Exploratory Spatial Data Analysis (ESDA) is a traditional method used to test spatiotemporal characteristics through spatial autocorrelation. Spatial autocorrelation is to explore spatial clustering or spatial anomalies by describing and visualizing the spatial pattern of a thing or phenomenon, which usually includes global spatial autocorrelation and local spatial autocorrelation. Global spatial autocorrelation is mainly used to analyze the spatial correlation of an element as a whole, which usually measured by Global Moran's I [40]. Its calculation equation is:

$$I = n\sum_{i=1}^{n}\sum_{j=1}^{n} w_{ij}(x_i - \overline{x})(x_j - \overline{x}) / \sum_{i=1}^{n}(x_i - \overline{x})\sum_{i=1}^{n}\sum_{j=1}^{n} w_{ij} \tag{3}$$

In Equation (3), n is the number of the study area; x is the CDUFS; $\overline{x}$ is its mean value; w_{ij} is the spatial weight matrix constructed from the reciprocal of geographical distance and normalized by rows, $\sum_j w_{ij} = 1$. $I \in [-1,1]$, $I > 0$, there is a spatial positive correlation in CDUFS; $I < 0$, there is a spatial negative correlation in CDUFS; $I = 0$, CDUFS is randomly distributed in space.

Local spatial autocorrelation reflects the cluster or outlier of attribute values between a region and its neighbors [41], and is used to observe the spatial imbalance, which is usually measured by Local Moran's I. The specific measure is LISA (Local Indicators of Spatial Association). Its calculation equation is:

$$I_i = z_i \sum\nolimits_j w_{ij} z_j \tag{4}$$

In Equation (4), z_i, z_j are the variance normalized values of CDUFS of city i, j, respectively.

LISA can classify CDUFS into four different types: High-High Cluster (H-H), where the CDUFS between a region itself and its neighboring regions are both high; Low-Low Cluster (L-L), where the CDUFS between a region itself and its neighboring regions are both low; High-Low Outlier, where the CDUFS in a region itself is high but its neighboring regions is low (H-L); Low-High Outlier (L-H), where the CDFUS in a region itself is low but its neighboring regions are high.

2.3.3. Regional Differences: Dagum–Gini Coefficient

The Dagum–Gini coefficient can analyze regional differences and decomposition of the CDUFS [42]. It can fully consider the distribution of sub-samples and decompose the overall Gini coefficient into intra-regional variation contribution, inter-regional variation contribution and transvariation density contribution [43], which can overcome the limitations of the traditional Gini coefficient and Theil index, effectively solving the issue of cross-overlap of sample data and the source of regional differences [44]. Before the calculation, the study object needs to be divided into different regions. The Dagum–Gini coefficient (G) for all regions is calculated as:

$$G = \sum_{j=1}^{k}\sum_{h=1}^{k}\sum_{i=1}^{n_j}\sum_{r=1}^{n_h}|y_{ji} - y_{hr}|/2n^2\overline{y} \tag{5}$$

In Equation (5), yji (y_{hr}) is the CDUFS of any region within j (h), $\overline{y}$ is the average of the CDUFS, n is the number of cities, k is the number of regions, j and h are region subscripts, i and r are the city subscripts.

In decomposing G by subgroups, first, the regions are ordered according to the mean value of the CDUFS, that is, $\overline{y_h} \leq \overline{y_j} \leq \ldots \leq \overline{y_k}$, then, the G is decomposed into three components: intra-regional variation contribution (G_w), inter-regional variation contribution (G_{nb}) and transvariation density contribution (G_i), which meet $G = G_w + G_{nb} + G_i$ [45]. The G portrays the size and source of the relative difference in the CDUFS.

$$\begin{gathered}G_w = \sum_{j=1}^{k} G_{jj}p_js_j,\ G_{nb} = \sum_{j=2}^{k}\sum_{h=1}^{j-1} G_{jh}(p_js_h + p_hs_j)D_{jh},\ G_t = \sum_{j=2}^{k}\sum_{h=1}^{j-1} G_{jh}(p_js_h + p_hs_j)(1 - D_{jh}),\\ G_{jj} = \frac{\frac{1}{2\overline{y}_j}\sum_{i=1}^{n_j}\sum_{r=1}^{n_j}|y_{ji}-y_{hr}|}{n_j^2},\ G_{jh} = \sum_{i=1}^{n_j}\sum_{r=1}^{n_h}\frac{|y_{ji}-y_{hr}|}{n_jn_h(\overline{y}_j+\overline{y}_h)},\ D_{jh} = \frac{d_{jh}-p_{jh}}{d_{jh}+p_{jh}},\\ d_{jh} = \int_0^{\infty} dF_j(y)\int_0^{y}(y-x)dF_h(x),\ p_{jh} = \int_0^{\infty} dF_h(y)\int_0^{y}(y-x)dF_j(x)\end{gathered} \tag{6}$$

In Equation (6), $p_j = n_j/n$, $s_j = n_j\overline{y_j}/n\overline{y}$, D_{jh} is the relative impact of the CDUFS in regions j and h. d_{jh} is the difference in the variation of the CDUFS among regions, which represents the mathematical expectation of all sample values sum of $y_{ji} - y_{hr} > 0$ in regions j and h. p_{jh} is the hypervariable first-order moment, which represents the mathematical expectation of all sample values sum of $y_{hr} - y_{ji} > 0$ in regions j and h. F_j (F_h) is the cumulative density distribution function of region j (h).

2.3.4. Dynamic Distribution: Kernel Density Estimation (KDE)

Kernel density estimation (KDE) is a nonparametric method that uses continuous density curves to describe the distribution of random variables. It can be used to analyze the distribution dynamics, polarization trends and distribution extension of CDUFS [46]. Suppose the density function of random variable x is $f(x)$, for x with n independent and identically distributed observations, $x_1, x_2, \ldots, x_n$, and $\overline{x}$ as its mean, the estimate of the Kernel density function is

$$f(x) = \frac{1}{nh}\sum_{i=1}^{n} K(\frac{x_i - \overline{x}}{h}) \tag{7}$$

In Equation (7), n is the study samples, h is the bandwidth, and $K(\cdot)$ is the stochastic kernel function, which is a weighting function or a smoothing transformation function, and the Gaussian (Normal) kernel function is used in this paper. The choice of bandwidth determines the smoothing degree of estimated density function. The optimal bandwidth must be chosen in a trade-off between the variance and bias of kernel estimation so that the mean square error is minimized.

2.3.5. Evolution Trend: Spatial Markov Chain

The Markov chain is a special kind of stochastic process, which can measure the state of an event occurrence and its development trend by constructing a state transfer probability matrix. The Markov chain follows the principle of "no aftereffect", and the conditional distribution of state X_t depends only on state X_{t-1} [47]. The evolution process of many economic phenomena such as the CDUFS possess "no aftereffect". We suppose that P_{ij} is the transition probability of the CDUFS from state i in year t to state j in year $t + 1$, and it can be estimated by transition frequency approximation, i.e., $P_{ij} = n_{ij}/n_i$. Where n_{ij} refers to the number of cities that transferred from state i to state j, and satisfy

$\sum_j P_{ij} = \sum_j P\{X_{n+1} = j | X_n = i\} = 1$. If the CDUFS is divided into N types, the state transition probability matrix can be constructed as $N \times N$. Moreover, the transition direction (downward, unchanged, upward) is defined according to the change of the CDUFS types.

The spatial Markov chain introduces spatial lag into the transition probability matrix, which makes up for the neglect of the spatial correlation effect in the traditional Markov chain and can be used to reveal the intrinsic connection between spatial-temporal evolution of the CDUFS and its spatial context. Taking the spatial lag type of a region in the initial year as the conditional basis, the spatial Markov chain can decompose the traditional $N \times N$ transfer probability matrix into $N \times N \times N$ transition probability matrix, so as to analyze the spatial dynamic evolution trend of the CDUFS and the impact of spatial effect under different geographical background conditions [48].

3. Results

3.1. Spatiotemporal Correlation and Differentiation

3.1.1. Global Spatial Autocorrelation

After measuring the CDUFS at the prefecture-level city, the national average CDUFS changed overall from 0.2965 to 0.2691, showing a significant downward trend and always fluctuating within the low coordination interval. The CDUFS showed an obvious regional differentiation in different food producing areas (Figure 2), mainly in the following ways: FPAs > FMAs > FBAs. Specifically, FPAs belonged to moderate coordination, while the FMAs and FBAs belonged to low coordination, and both showed a fluctuating downward trend, with the FMAs showing a more prominent downward trend.

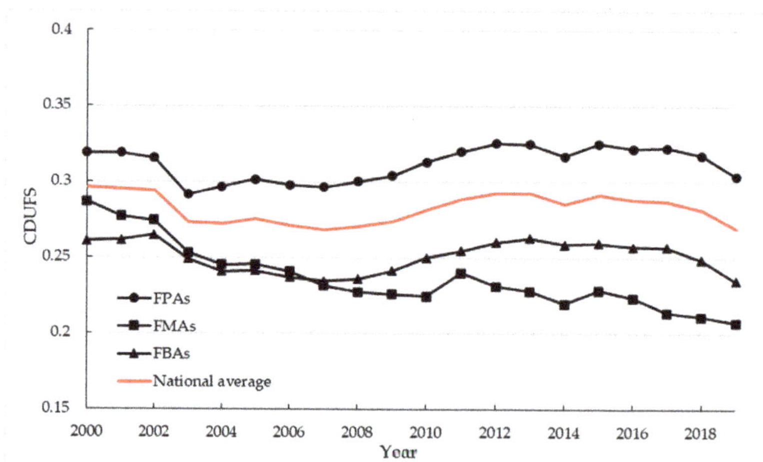

Figure 2. Time variation of CDUFS in different food functional areas. Note: This figure is drawn by the authors themselves based on the calculation results.

Related studies have confirmed that the spatial mobility of factors makes a significant spatial correlation to both urbanization and food production [49,50]. Based on the distance relationship between regions, Moran's I for the CDUFS from 2000–2019 was calculated. The Moran's I $\in$ [0.353, 0.674] of the CDUFS at the prefecture-level city all passed the significance test at the 1% level (Table 2). Moran's I basically showed a steady increasing trend. The CDUFS had a significant positive spatial correlation and dependence, and this positive correlation in space showed a stable and continuous strengthening.

Table 2. Moran's I of the CDUFS.

Year	Moran's I	Z	Year	Moran's I	Z	Year	Moran's I	Z	Year	Moran's I	Z
2000	0.473	13.615	2005	0.623	17.904	2010	0.655	18.812	2015	0.652	18.734
2001	0.353	10.189	2006	0.617	17.74	2011	0.587	16.894	2016	0.654	18.782
2002	0.483	13.914	2007	0.603	17.337	2012	0.669	19.223	2017	0.667	19.156
2003	0.452	13.044	2008	0.637	18.292	2013	0.665	19.099	2018	0.674	19.358
2004	0.607	17.449	2009	0.621	17.847	2014	0.642	18.452	2019	0.649	18.670

3.1.2. Local Spatial Autocorrelation

To further portray the spatial differentiation of the CDUFS, the years 2000, 2005, 2010, 2015 and 2019 were selected for LISA analysis (Figure 3). The FPAs in the Northeast Plain and the Huang-huaihai Plain showed H-H significantly. The FBAs or FMAs in the Northwest, the Southwest and the South China coast showed significant L-L, while showing H-L and L-H sporadic distribution around H-H and L-L, but fluctuations also occurred in local areas.

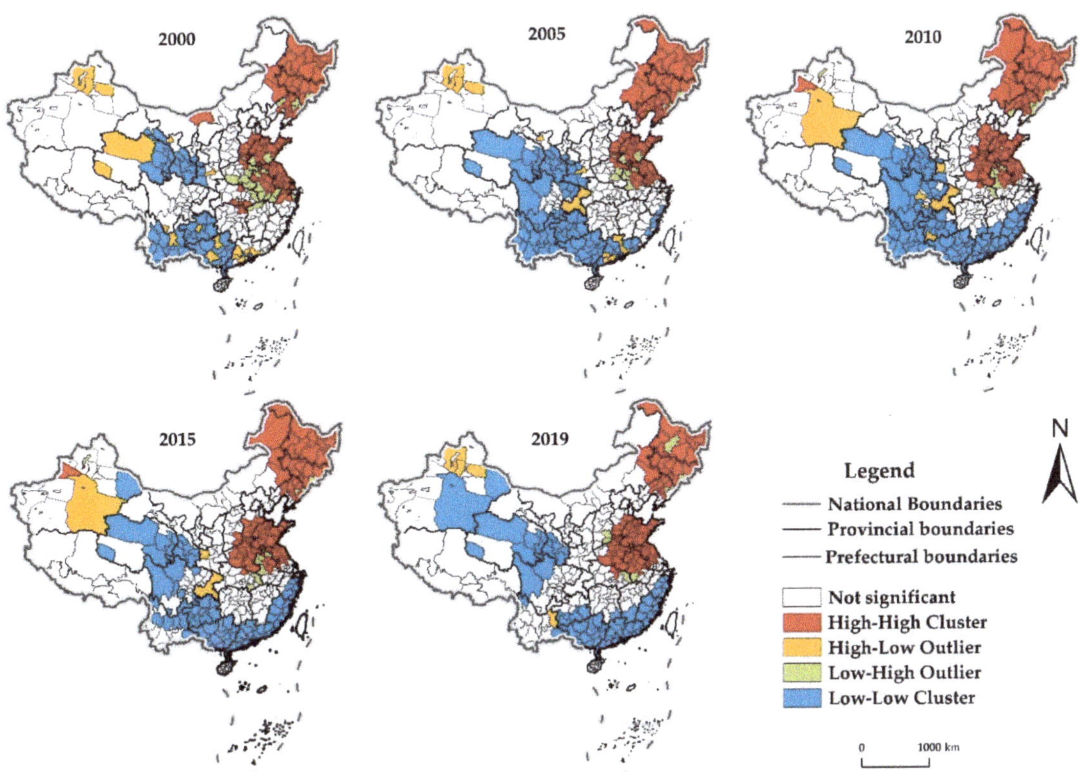

Figure 3. LISA distribution of CDUFS. Note: This figure is drawn by the authors themselves using ArcGIS based on the calculation results.

(1) High-High Cluster (H-H). Areas with H-H are mainly distributed in the FPAs such as the Northeast Plain, the North China Plain, and the middle and lower reaches of the Yangtze River. The coverage of H-H in general also showed a steady expansion of changes. The H-H areas in 2000 were mainly located in the FPAs of Northeast Plain, North China Plain and the lower Yangtze River. By 2019, the distribution expanded significantly to the FPAs in the center region, especially Hubei and Henan. Overall,

the distribution of H-H was more stable, and its distribution range was gradually expanding to the FPAs in the middle reaches of the Yangtze River.

(2) Low-Low Cluster (L-L). Areas with L-L were steadily at the Northwest, Southwest and Southeast coast, concentrated in the FBAs and FMAs. In terms of changes in their coverage, it showed a stable expansion in the Northwest, expansion followed by contraction in the Southwest, and obvious expansion in the Southeast coastal. In general, the distribution pattern of H-H was relatively stable, but the southwest region shrunk significantly, and the distribution range gradually expanded to the FMAs of the Southeast coast.

(3) High-Low Outlier and Low-High Outlier (H-L, L-H). Areas with H-L and L-H of CDUFS were less distributed and sporadic, concentrated in the peripheral neighboring areas of H-H and L-L. The interaction between areas weakened the CDUFS to a certain extent. Individual areas in Qinghai have changed from H-L to L-L, which means that its urbanization and food security are faced with great challenges. Other areas with H-L are mainly at Xinjiang, Yunnan, Guangxi, etc., concentrated in the periphery of L-L. The areas with H-L in Yunnan and Guangxi are gradually disappearing while areas with L-H include the Daxinganling in Heilongjiang, mainly at the periphery of H-H.

3.2. *Regional Differences and Decomposition*

Although the analysis of spatiotemporal correlation and differentiation can visually portray the heterogeneous distribution of the CDUFS among regions, it is difficult to further reflect the extent of regional differences and their causes. Therefore, the Dagum–Gini coefficient method was used to systematically analyze the overall differences in the CDUFS, the differences in different food functional areas and their sources.

3.2.1. Overall Regional Differences

The Dagum–Gini coefficient of CDUFS ranged from 0.0953 to 0.1360, with a mean value of 0.1223 during study period, implying that the coordination relationship between urbanization and food security exhibited a certain spatial disequilibrium. After a brief oscillation until 2004, the Gini coefficient of the the CDUFS showed an upward trend and reached a maximum in 2011 (0.1360), after which it declined slightly in fluctuations (Figure 4). The Gini coefficient of the CDUFS increased by 0.0305, or 32.00%, with a more significant increase, so the overall regional differences of the CDUFS in China have expanded.

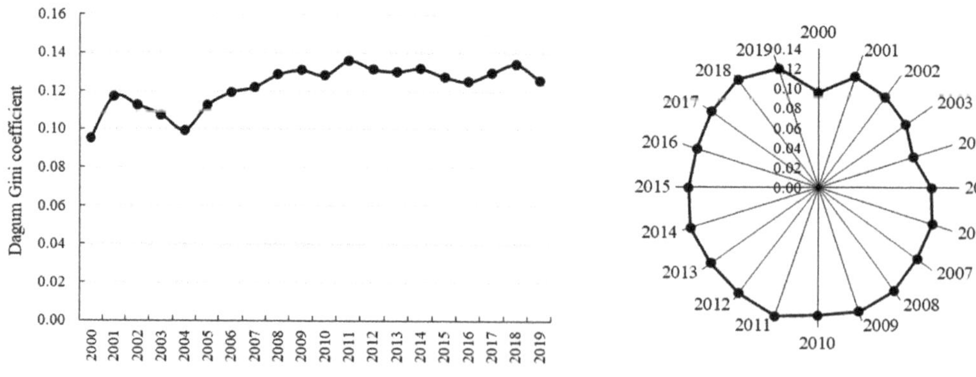

Figure 4. Overall regional differences in the CDUFS. Note: This figure is drawn by the authors themselves based on the calculation results.

3.2.2. Regional Differences and Their Decomposition

For intra-regional differences (Table 3), the mean values of the Gini coefficient of the CDUFS in the FPAs, FMAs and FBAs were 0.0933, 0.0958 and 0.1181, respectively, and the internal differences of coupling coordination in FMAs were higher than those in the FPAs and FBAs. The intra-regional differences of the CDUFS in the three food functional areas had different trends, but compared with 2000, they all showed an overall increase, indicating that the internal differences of different food functional areas of CDUFS have expanded. The Gini coefficient of FPAs changed from 0.0742 in 2000 with an increasing and then decreasing trend to 0.0774 in 2004, and then gradually increased steadily to 0.0882 in 2019, with an overall increase of 18.87%. The Gini coefficient of FMAs, except for the anomaly in 2011 (0.1433), basically showed a steady increase, from 0.0750 to 0.1126 year-on-year, an overall increase of 50.13%. The Gini coefficient of FBAs changed from 0.0959 in 2000 to 0.0912 in 2004 first increasing and then decreasing, then increased to 0.1003 in 2019, with a relatively small increase of 4.59% overall. In addition, the expansion trend of internal differences of the CDUFS was the most pronounced in FMAs, followed by FPAs.

Table 3. Regional differences and their sources of CDUFS in different functional areas.

Food Functional Areas		2000	2003	2006	2009	2012	2015	2018	2019	Mean
Intra-regional difference	FPAs	0.0742	0.0848	0.0950	0.0983	0.1008	0.0965	0.0948	0.0882	0.0933
	FMAs	0.0750	0.0812	0.0838	0.0956	0.0968	0.0992	0.1116	0.1126	0.0958
	FBAs	0.0959	0.1255	0.1217	0.1352	0.1261	0.1133	0.1185	0.1003	0.1181
Inter-regional differences	FPAs-FMAs	0.0860	0.1000	0.1254	0.1602	0.1772	0.1814	0.2065	0.1938	0.1538
	FPAs-FBAs	0.1210	0.1293	0.1449	0.1540	0.1475	0.1420	0.1481	0.1429	0.1414
	FMAs-FBAs	0.0970	0.1087	0.1060	0.1200	0.1211	0.1177	0.1334	0.1186	0.1157
Contribution rate/%	Intra-regional	35.42	37.15	36.35	35.18	35.26	34.44	32.92	32.07	35.07
	Intra-regional	47.55	35.25	46.16	48.22	51.14	53.78	57.11	60.11	49.40
	Transvariation density	17.03	27.61	17.49	16.60	13.60	11.78	9.97	7.82	15.53

For inter-regional differences, the mean value of the Gini coefficient of the CDUFS was ranked mainly FPAs-FMAs (0.1538), FPAs-FBAs (0.1414), FMAs-FBAs (0.1157), so the difference of the coupling coordination between FPAs and FMAs were higher than those between other functional areas. The Gini coefficient of FPAs-FMAs showed a stable and continuous upward trend, from 0.0860 in 2000 to 0.1938 in 2019, with an obvious increase of 1.25 times, which further highlights the spatial non-equilibrium between FPAs and FMAs. The Gini coefficient of FPAs-FBAs had a relatively stable variation, increasing from 0.1210 in 2000 to 0.1429 in 2019, an overall increase of 18.10%. The Gini coefficient of FMAs-FBAs was the smallest in most years, with small differences between regions, but its overall growth trend is slightly fluctuating, increasing from 0.0970 to 0.1186 year-on-year, an increase of 22.27%. In contrast, the inter-regional difference of CDUFS between FPAs and FMAs showed a more prominent trend of expansion.

For the sources and contributions of regional difference, the mean contributions of intra-regional difference, inter-regional difference and transvariation density were 35.07%, 49.40% and 15.53%, respectively, showing that the inter-regional difference was the main source of overall difference in functional areas with a contribution of nearly 50%, followed by an intra-regional difference of functional areas and the smallest contribution of transvariation density. The contribution of inter-regional difference is in the range of 35.25% to 60.11%, basically showed a stable growth trend with an overall increase of 26.41%, except for a slight fluctuation before 2004. The contribution of intra-regional difference was in the range of 32.07% to 38.88%, showing a decline trend of first rising and then falling, but the volatility was relatively stable, with an overall decline of 9.46%. The contribution of transvariation density was in the range of 7.82% to 27.61%. Its changes were basically opposite to the inter-regional difference in a steady downward trend, with an overall decrease of 9.21 percentage points, indicating that the contribution of the cross-overlapping effect on the CDUFS between intra-regional and inter-regional to overall difference is gradually decreasing.

3.3. Dynamic Distribution

To further clarify the dynamic distribution of the CDUFS, the KDE method was used to portray the distribution location, distribution pattern, distribution ductility and polarization characteristics of the CDUFS at the national level and different food functional areas (Figure 5).

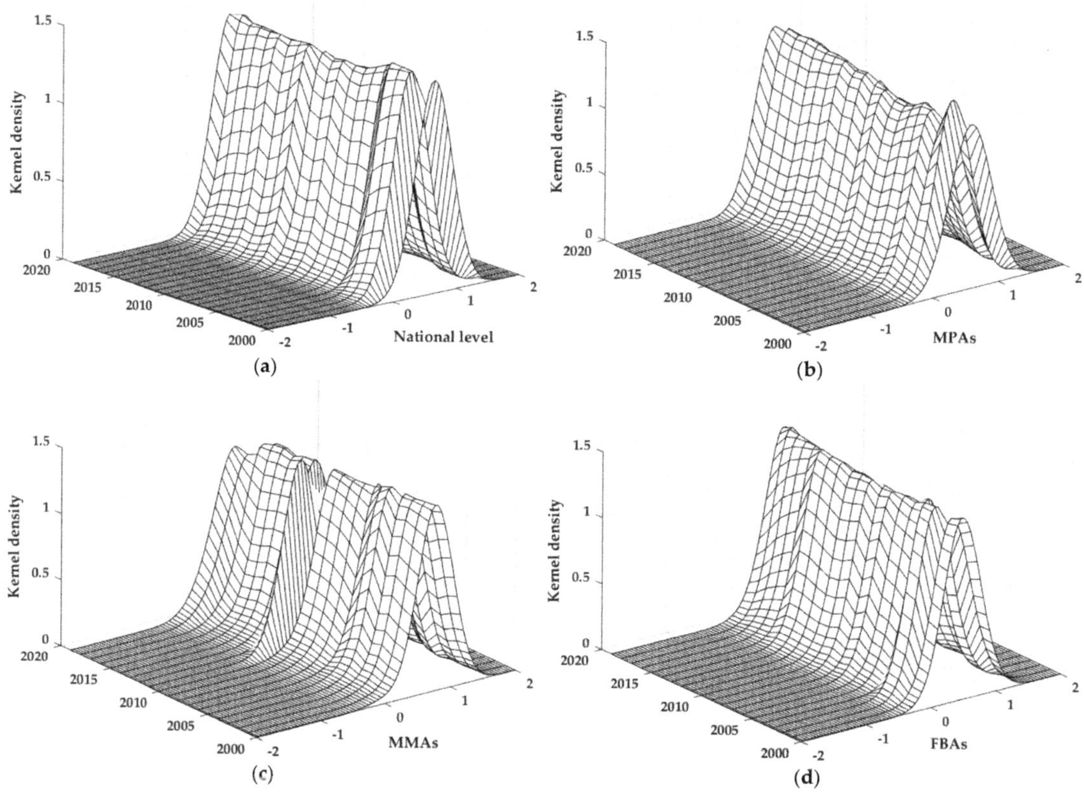

Figure 5. Kernel density estimation of the national level and different food functional areas. Note: This figure is drawn by the authors themselves using MATLAB based on the calculation results. (**a**) National level; (**b**) FPAs; (**c**) FMAs; (**d**) FBAs.

3.3.1. The National Level

(1) Distribution location. The Kernel density curve at the national level showed an evolution process in which the curve center first shifted left and then right, and the peak height first increased and then decreased, indicating that the CDUFS generally showed a decreasing trend over time.

(2) Distribution pattern. The Kernel density curve showed a significant single-peak distribution with a relatively stable distribution trend. On the whole, the width of the main peak has increased, and the height of the main peak has decreased, indicating that the imbalance phenomenon in the CDUFS is still prominent, and

the coordination gap between regions tends to expand, which is consistent with the typical characteristics in the previous section.

(3) Distribution ductility. The Kernel density curve had a weak left trailing phenomenon, and the distribution ductility was broadened to a certain extent but did not show a convergence trend. The gap of the CDUFS between the regions with high coupling coordination and the average level was somewhat enlarged.

(4) Polarization characteristics. The distribution of the Kernel density curve with a single-peak and a weak left trailing, indicating that the CDUFS did not show a polarization state and gradient effect, the single-peak distribution would continue to exist, and the gap of CDUFS between regions was more obvious.

The development of urbanization and food security have both made great progress to different degrees, but urbanization, which relies on economic growth, is growing faster, and food production still lags behind the urbanization process. In addition, there is heterogeneity in the endowment conditions and development environment among different regions. These are the realistic reasons for the single-peak distribution of CDUFS.

3.3.2. Different Food Functional Areas

(1) Distribution location. The centers of Kernel density curve in both FPAs and FBAs evolved leftward and then rightward, with an overall shift to the left, similar to the national level, while the center of Kernel density curve in FMAs showed an obvious leftward evolution, indicating that the CDUFS in different food functional areas have a decreasing trend over time.

(2) Distribution pattern. The Kernel density curves of the three food functional areas were all significant single-peak patterns, with the width of the main peak increasing and the height of the main peak decreasing. Among them, the main peak height of FPAs showed a continuous decreasing process, while the main peak heights of FMAs and FBAs experienced a rising and then decreasing process, which generally indicates that the spatial imbalance of the CDUFS in different food functional areas is still prominent. The trends for this coordination gap widened.

(3) Distribution ductility. The Kernel density curve of FPAs did not show an obvious trailing phenomenon, while FMAs and FBAs showed left trailing phenomenon, and the FMAs were more prominent with a certain widening of the distribution ductility. The Kernel density curves of the three food functional areas did not show a convergence trend, and the gap between the areas with high coordination and the average level widened, prominently in the FMAs.

(4) Polarization characteristics. The Kernel density curves of different food functional areas with a single-peak indicates that the CDUFS of each food functional area did not have a polarized state and gradient effect, the single-peak distribution will continue to exist, and the coordination gap between regions is still obvious.

3.4. Evolution Trend

Although the Dagum–Gini coefficient and KDE can portray the regional differences and dynamic distribution of the CDUFS, it is difficult to deeply reflect the dynamic evolution trend of this spatiotemporal difference. So, the traditional and spatial Markov chain analysis were used to deeply examine the evolution trend of the CDUFS.

The basis of the Markov chain analysis is the transition probability matrix. It was necessary to divide the CDUFS of 330 prefecture-level and above cities during 2000–2019 into different types of state spaces. Based on the quantile division method [48], using the 1/4, 1/2 and 3/4 quantile as the boundaries, the CDUFS was divided into four adjacent but non-crossing completeness intervals: [0.0150, 0.2406], (0.2406, 0.2777], (0.2777, 0.3236], (0.3236, 0.5663], the completeness intervals of these four state types can be represented by $k = 1, 2, 3$, and 4, respectively. The larger k is, the higher CDUFS is (lower, low, high, higher).

3.4.1. Traditional Markov Chain Analysis

The traditional Markov transition probability matrix was calculated based on the division of state types (Table 4). The elements on the diagonal line represent the probability that the state type of the CDUFS does not transfer in a region, which reflects the stability of the evolution trend of the CDUFS in this region, while the elements on the non-diagonal line represent the probability that the CDUFS in a region transfers between different state types. Therefore, without considering the spatial effect, the evolution of the CDUFS is characterized as follows:

(1) The CDUFS had the stability of keeping the original state. All the elements on the diagonal were larger than those on the non-diagonal significantly, implying that the CDUFS had a smaller probability of state transfer. The fluidity among states was low with an obvious path-dependence. In addition, the probability of maintaining stability was greatest for the types at either end of the diagonal (types 1 and 4), while types 2 and 3 in the diagonal had a relatively smaller probability of maintaining stability. The CDUFS was more likely to be distributed at lower and higher levels, low and high distributions more prone to transfer.

(2) Areas with a higher level of CDUFS were more likely to transfer to the lower level, while areas with a lower level of CDUFS had a higher likelihood of shifting to a higher level. For example, the probability of type 4 transition downward was 0.1091, the probability of type 3 transition downward was 0.1073, the probability of type 2 transition downward was 0.1013, 0.1091 > 0.1073 > 0.1013. It can be seen that areas in type 4 were more likely to transfer downward. Although the evolution trend was stable in areas with a high level of CDUFS, there is a certain risk and possibility of falling back, which requires vigilance and attention. In addition, the probability of type 2 transferring upward was greater than that of type 3 (0.1742 > 0.1171), indicating that areas with a relatively low CDUFS have higher room for growth.

(3) The CDUFS was difficult to achieve a leapfrog evolution in the short term. The probability transition of the CDUFS occurred almost on both sides of the diagonal. For elements on both sides of the non-diagonal, the probability values were significantly smaller than those on both sides of the diagonal. For example, the probability of type 2 transferring upward to type 4 was 0.0083, which is obviously smaller than the probability of transferring to type 3 (0.1389). Among two consecutive years, the probability of achieving a leapfrog transfer was low (e.g., 1→3, 1→4), which means that the evolution of CDUFS in each region was a relatively stable and continuous process.

Table 4. Traditional Markov chain transition probability matrix for CDUFS.

Type	n	1	2	3	4
1	1614	0.8482	0.1363	0.0105	0.0050
2	1570	0.1013	0.7516	0.1389	0.0083
3	1537	0.0078	0.0995	0.7755	0.1171
4	1549	0.0039	0.0077	0.0975	0.8909

3.4.2. Spatial Markov Chain Analysis

The probability transfer matrix of the traditional Markov chain did not consider the spillover effect of type of transfer in neighboring regions, but the CDUFS had a significant positive correlation and dependence in space. Therefore, the type of transfer of the CDUFS is not isolated in space but is complementary and effectively linked to the surrounding areas. Introducing the spatial lag effect, the spatial Markov chain transition probability matrix (Table 5) was constructed based on the spatial lag type of each area in the initial year. In addition to the common features with the traditional Markov chain transition probability matrix, it also had the following spatial evolution features.

(1) Geospatial pattern plays an important role in the dynamic evolution of the CDUFS, and spatial Markov chain analysis can provide a spatially meaningful interpretation of the single-peak distribution pattern of the CDUFS. Under the spatial effect, the type of transition probability of the CDUFS in each area was not the same, was also not equal to the corresponding traditional Markov transition probability matrix, otherwise, the effect of spatial lag would not exist (Figure 6). For example, when the spatial effect was not considered, the transition probability from type 2 to type 3 in an area was $P_{23} = 0.1389$. When the area was adjacent to an area in type 2, $P_{23|2} = 0.1019$, when it was adjacent to an area in type 3, $P_{23|3} = 0.1529$, and when it was adjacent to an area in type 4, $P_{23|4} = 0.2237$. It can be seen that it is necessary to consider the spatial background when analyzing the spatial evolution of the CDUFS. An area is adjacent to the areas of different types, the state transition probability of the CDUFS in this area will be different. In general, for a certain area, the probability of upward transfer of its type was greater when adjacent to areas with higher coordination, while the probability of downward transfer of its type was greater when adjacent to an area with lower coordination.

(2) The evolution of the CDUFS remained more stable in its original state; the possibility of jump transfer is low. This is because the elements on the diagonal were still larger than the elements on the non-diagonal after considering the spatial background. As the spatial lag type rises, there are differences in the stability and transition probability of the evolution of the CDUFS. For areas with a lower CDUFS (type 1), their stability decreased as the spatial lag type increased, but the probability of upward transfer was increasing. For areas with a low CDUFS (type 2), their stability showed an increase and then decrease with the increase of spatial lag type, while the probability of upward transfer showed a decrease and then increase, and the probability of downward transfer in decreasing. For areas with a high CDUFS (type 3), their stability also showed an increase and then a decrease with the increase of spatial lag type, but the probability of upward transfer was increasing and the probability of downward transfer was decreasing. For areas with a higher CDUFS (type 4), their stability decreased and then increased with the increase of the spatial lag type, while the probability of downward transfer showed an increase and then a decrease.

Table 5. Spatial Chain Transition Probability Matrix for CDUFS.

Spatial Lag	Type	n	1	2	3	4
1	1	250	0.9120	0.0840	0.0040	0
	2	116	0.1121	0.7586	0.1293	0
	3	41	0.0244	0.2683	0.6829	0.0244
	4	11	0	0	0.0909	0.9091
2	1	736	0.8641	0.1209	0.0095	0.0054
	2	589	0.1121	0.7827	0.1019	0.0034
	3	233	0.0258	0.1760	0.7425	0.0558
	4	95	0.0211	0.0211	0.1263	0.8316
3	1	595	0.8151	0.1681	0.0118	0.0050
	2	713	0.1010	0.7377	0.1529	0.0084
	3	833	0.0060	0.0960	0.8019	0.0960
	4	595	0.0050	0.0118	0.1193	0.8639
4	1	33	0.6061	0.3030	0.0606	0.0303
	2	152	0.0526	0.6908	0.2237	0.0329
	3	430	0	0.0488	0.7512	0.2000
	4	848	0.0012	0.0035	0.0790	0.9163

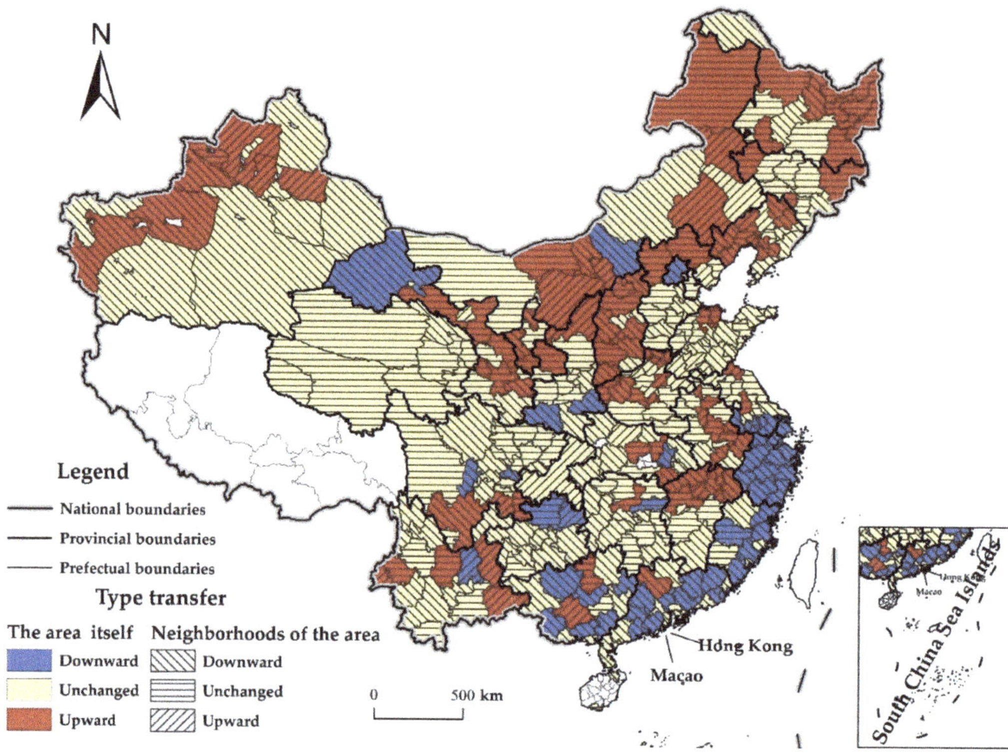

Figure 6. Spatial distribution of CDUFS type of transfer under the spatial Markov chain. Note: This figure is drawn by the authors themselves using ArcGIS based on the calculation results.

4. Discussion

While previous studies have focused more on the one-way impact of urbanization on food security and the challenges faced by food security in the urbanization process [51], this paper innovatively explores the interaction between urbanization and food security. The CDUFS always fluctuates in the low coordination interval, which is different from the study of Yao [20]. Both urbanization and food security have achieved stable growth, but the interaction effect between the two has not achieved synchronous growth. It mainly stems from the fact that the growth of food security continues to lag behind urbanization development, and the unbalanced development of the two constrains the improvement of coupling coordination level. In addition, the distribution pattern of the agglomeration of the CDUFS basically keeps a steady trend and shows a "center-periphery" with FPAs in the north gradually decreasing to FBAs in the northwest and FMAs in the south.

The CDUFS of China show obvious regional differentiation, and the overall regional differences of the CDUFS in China have expanded. The possible explanation is that, although food security is increasingly guaranteed, the growth process of urbanization is still faster than the supply capacity of food production. It has been found that urbanization inevitably brings about the reallocation between urban and rural areas [22], and the regional differences in endowment conditions and industrial structure also make the relationship between urbanization and agricultural production exist in regions located in different geographical spaces. Continued urbanization and changing consumption patterns pose challenges to food security. Avoiding deterioration in food security depends on the responsiveness and resilience of the smallholder farming sector. Rural-urban food supply linkages [52] and policy synergies are also critical.

For regional differences in different food functional areas, inter-regional differences are the main source of overall differences, mainly in the widening trend of differences in coupling coordination between FPAs and FMAs, and the internal differences of FMAs are also expanding. The FPAs face the dual pressure of ensuring food security and urbanization development, and the supply task of food production in FPAs is equally important in the urbanization process, while FMAs basically cover the more economically developed areas of East China and the South China coast. The industrial structure upgrading implies the proportion of agriculture is declining. There is the adjustment of structure in food production to make room for urbanization development, and the coordination level between urbanization and food security is poor.

The regional differences between FPAs and FMAs make us alert to a mismatch between the main body of food production and the spatial distribution of soil-water resources in the South and North [53]. The production and marketing pattern of "northern food transportation to the south" remains stable. The main producing areas undertake more food supply tasks, but their soil-water conditions do not have resource advantages, and the gap between grain supply and demand is gradually increasing in some northern provinces of main producing areas. Therefore, it is possible to explore the pilot mechanism of interest linkage between the main producing areas and the main marketing areas. Through the pilot project, according to their own development base and the total amount of grain transferred, the main marketing areas in the southern area can compensate for the benefits of the main producing areas in the northern area, promote agricultural investment in the main producing areas in order to guarantee grain production in the main producing areas, and maintain a long-term stable inter-regional relationship between grain supply and demand, so that the main marketing areas can also assume a certain responsibility for ensuring food security in the form of trade.

Amongst the several challenges, food security will be a serious issue for the future of high-quality urbanization [54]. This study is of great significance for understanding the coordinated relationship between urbanization and food security, exploring a sustainable path for high-quality urbanization and guaranteeing food security in China, and promoting a synergistic balance between urbanization and agricultural production.

(1) Changes in spatial use of territory land led to the conflict between urbanization and food security. It is necessary to plan the urban-rural land use scientifically and reasonably, clarify the limit of urban land, and slow down the expansion speed of urban land. In addition, we should strengthen the policy of farmland protection in the urbanization process, strictly guard the red line of cropland use, and optimize the balance management system of farmland occupation and supplement to maintain the dynamic balance of total farmland, so as to guarantee the high-quality development of urbanization and sustainable food production.
(2) Urbanization not only leads to the adjustment in production resources and grain-growing behavior, but also provides an opportunity for large-scale, mechanization and efficient operation of food production [55]. Therefore, the process of high-quality urbanization involves guiding households to transfer management rights of land orderly to bring into play the scale effect of land. Improving food production efficiency and yield capacity through intensive land use and technological progress can also provide useful support for urbanization, which promote the coordinated development between urbanization and food security.
(3) Differences in industrial division of labor make regional differences in the CDUFS, and there is no convergence in the long-term evolution process. Therefore, the strategies of urbanization development and ensuring food security by different regions should be adapted to local conditions, and in line with their own development positioning based on industrial characteristics.
(4) The FPAs need to grasp the room and direction in the process of urbanization, avoid the irregular expansion of urban space, ensure the free mobility of factors between urban areas and rural areas, adapt the utilization structure of water-soil resources,

enhance the efficiency of land use, and drive the intensive management of food productive through urbanization. The FMAs have a high level of urbanization, and with the industrial structure upgrading, the comparative advantage of agricultural production has declined obviously. It should focus on weighing the intrinsic structure of food production against the extrinsic changes triggered by urbanization and improve the factor agglomeration and labor productivity through technological progress. The urbanization of FBAs is relatively backward, but grain producing and marketing basically maintain a balance. Therefore, the potential for coordination between urbanization and food security is greater. The trade-off between the government's guiding effect and the market's allocation effect should be focused on to optimize the regional layout of urbanization and food production by effective cooperation between government and market.

Based on the panel data of prefecture-level, this study allowed for a more detailed exploration on a smaller dimension. However, due to the difficulty of obtaining data on relevant indicators, food security is represented only as a single indicator. More complete data acquisition is what needs to be improved in subsequent research.

5. Conclusions

This paper measures the CDUFS in China using prefecture-level city panel data and the coupled coordination degree model based on understanding of the spatial correlation and differentiation of the CDUFS by ESDA. From the perspective of different food functional areas, regional differences, and sources, distribution dynamics and evolutionary trends of the CDUFS were analyzed in depth using the Dagum–Gini coefficient, the Kernel density estimation and the spatial Markov chain. The main findings are as follows:

(1) The CDUFS in China showed a downward trend in fluctuating within the low coordination range, and food security continues to lag behind urbanization. The regional differences of the CDUFS are obvious with a continuously enhanced positive correlation in space. This exhibits a stable distribution pattern of H-H in the Northeast Plain and Huang-Huaihai Plain, and L-L in the Northwest, Southwest and Southeast coast.

(2) There were obvious regional differences in the CDUFS in China, and the overall differences have expanded. For different food functional areas, inter-regional differences were the main source of contribution to the overall differences, with the highest regional differences between FPAs and FMAs, and their spatial imbalance gradually prominent. The contribution of intra-regional differences was the second, with the highest of internal differences in FBAs, and its expanding trend of internal differences was more obvious. The contribution of transvariation density was the smallest.

(3) The main peak of the distribution curve of the CDUFS in China has increased in width and decreased in height. The imbalance of the CDUFS between regions is still prominent, but it does not show a convergence trend, and the single-peak distribution will continue to exist. The CDUFS in the three food function areas show a significant leftward shift of the single-peak distribution process, and the distribution curve of the FPAs do not show a trailing phenomenon. While the FMAs and FBAs show the left trailing, the FMAs are more obvious.

(4) The CDUFS in China have the stability of maintaining the original state, obviously path dependent. The trend of the CDUFS transferring to a high level is not obvious, but areas with higher CDUFS have a higher possibility of transferring to a low level. It is difficult to achieve a leapfrog evolution of the in the short term. Geospatial pattern plays an important role in the dynamic evolution of CDUFS, and there are differences in the stability and transfer probability of the CDUFS evolution under the different spatial lag type. The long-term evolution trend of CDUFS in China is influenced by the geospatial effect, but the single-peak pattern is relatively stable and does not exhibit significant convergence in the long term.

Author Contributions: Conceptualization, S.L. and M.H.; methodology, S.L.; software, M.H.; validation, S.L. and M.H.; formal analysis, M.H.; investigation, S.L.; resources, M.H.; data curation, M.H.; writing—original draft preparation, S.L.; writing—review and editing, M.H.; visualization, M.H.; supervision, S.L. and M.H.; project administration, S.L.; funding acquisition, M.H. All authors have read and agreed to the published version of the manuscript.

Funding: This research was funded by the Scientific Research Initiation Project for High-level Talents of Hebei University, grant No. 521100222017.

Institutional Review Board Statement: Not applicable.

Informed Consent Statement: Not applicable.

Data Availability Statement: The data presented in this study are available on request from the corresponding author. The data are not publicly available due to data management.

Conflicts of Interest: The authors declare no conflict of interest.

Abbreviations

CDUFS	coupled coordination degree between urbanization and food security
FPAs	food main producing areas
FMAs	food main marketing areas
FBAs	food balanced areas

References

1. Pecl, G.T.; Araújo, M.B.; Bell, J.D.; Blanchard, J.; Bonebrake, T.C.; Chen, I.C.; Clark, T.D.; Colwell, R.K.; Danielsen, F.; Evengård, B.; et al. Biodiversity redistribution under climate change: Impacts on ecosystems and human well-being. *Science* **2017**, *355*, eaai9214. [CrossRef] [PubMed]
2. Liu, Y. Is the natural resource production a blessing or curse for China's urbanization? Evidence from a space–time panel data model. *Econ. Model.* **2014**, *38*, 404–416. [CrossRef]
3. Song, H. Food demand-supply balance and state food safety. *J. Nanjing Agric. Univ.* **2016**, *16*, 1–11.
4. Tian, X.; Yi, F.; Yu, X. Rising cost of labor and transformations in grain production in China. *China Agric. Econ. Rev.* **2020**, *12*, 158–172. [CrossRef]
5. Rosenzweig, C.; Mbow, C.; Barioni, L.G.; Benton, T.G.; Herrero, M.; Krishnapillai, M.; Liwenga, E.T.; Pradhan, P.; Rivera-Ferre, M.G.; Sapkota, T.; et al. Climate change responses benefit from a global food system approach. *Nat. Food* **2020**, *1*, 94–97. [CrossRef]
6. Zhang, Q.; Zhang, F.; Wu, G.; Mai, Q. Spatial spillover effects of grain production efficiency in China: Measurement and scope. *J. Clean. Prod.* **2021**, *278*, 121062. [CrossRef]
7. Branca, G.; Lipper, L.; Mccarthy, N.; Jolejole, M.C. Food security, climate change, and sustainable land management, a review. *Agron. Sustain. Dev.* **2013**, *33*, 635–650. [CrossRef]
8. Liu, Y.; Zhou, Y. Reflections on China's food security and land use policy under rapid urbanization. *Land Use Policy* **2021**, *109*, 105699. [CrossRef]
9. Fan, Q.; Qi, H. Transforming urbanization development and ensuring national food security. *Macroeconomics* **2014**, *8*, 54–60. [CrossRef]
10. Yue, W.; Liu, Y.; Fan, P. Measuring urban sprawl and its drivers in large Chinese cities: The case of Hangzhou. *Land Use Policy* **2013**, *31*, 358–370. [CrossRef]
11. Leng, Z.; Fu, C. The influence of the disequilibrium development of urbanization on food security. *Economist* **2014**, *11*, 58–65. [CrossRef]
12. Chen, J. Rapid urbanization in China: A real challenge to soil protection and food security. *CATENA* **2007**, *69*, 1–15. [CrossRef]
13. Ferrara, A.; Salvati, L.; Sabbi, A.; Colantoni, A. Soil resources, land cover changes and rural areas: Towards a spatial mismatch? *Sci. Total Environ.* **2014**, *478*, 116–122. [CrossRef] [PubMed]
14. Li, Q.; Xia, M.; Guo, X.; Shi, Y.; Lu, H. Spatial characteristics and influencing factors of risk perception of haze in China: The case study of publishing online comments about haze news on Sina. *Sci. Total Environ.* **2021**, *785*, 147236. [CrossRef] [PubMed]
15. He, C.; Liu, Z.; Xu, M.; Ma, Q.; Dou, Y. Urban expansion brought stress to food security in China: Evidence from decreased cropland net primary productivity. *Sci. Total Environ.* **2017**, *576*, 660–670. [CrossRef] [PubMed]
16. Seto, K.; Ramankutty, N. Hidden linkages between urbanization and food systems. *Science* **2016**, *352*, 943–945. [CrossRef]
17. Lu, W.; Chen, N.; Qian, W. Modeling the effects of urbanization on grain production and consumption in China. *J. Integr. Agric.* **2017**, *16*, 1393–1405. [CrossRef]
18. Xu, J.; Zha, T. Provincial food security from the perspective of urbanization in Jiangsu province. *Resour. Sci.* **2014**, *36*, 2353–2360.
19. Gao, Y.; Zhang, Z.; Wei, S.; Wang, Z. Impact of urbanization on food security: Evidence from provincial panel data in China. *Resour. Sci.* **2019**, *41*, 1462–1474. [CrossRef]

20. Yao, C.; Qiu, Y. Coupling relationship between urbanization and food security in China: An empirical study. *China Soft Sci.* **2016**, *8*, 75–88.
21. Zhu, Y.; Wang, Z.; Luo, J.; Cui, J. Coupling and coordinated development of new urbanization and food security in main agricultural production regions in central China: A case study of Henan Province. *Sci. Geogr. Sin.* **2021**, *41*, 1947–1958. [CrossRef]
22. Hou, M.; Deng, Y.; Yao, S. Coordinated relationship between urbanization and grain production in China: Degree measurement, spatial differentiation and its factors detection. *J. Clean. Prod.* **2022**, *331*, 129957. [CrossRef]
23. Ren, X.; Wu, Y.; Wang, M. Study on the dynamic relationship between urbanization rate and cultivated land pressure in China from 1998-2018. *Chin. J. Agric. Resour. Reg. Plan.* **2022**, *43*, 120–130.
24. Zhao, L.; Hou, D.; Wang, Y.; Yi, N. Study on the interactive relationship between urbanization and the technical efficiency of grain production. *China Popul. Resour. Environ.* **2017**, *27*, 106–114. [CrossRef]
25. Gong, R.; Xie, L.; Wang, Y. Interactive mechanism and empirical test of agricultural high-quality development and new urbanization. *Reform* **2020**, *7*, 145–159.
26. Zeng, F.; Gao, M. Coordinated development of agricultural modernization, industrialization and urbanization in China and its influencing factors–Based on modern agriculture perspective. *Chin. Rural. Econ.* **2013**, *1*, 24–39.
27. Li, B.; Kong, X. A study on the pulling effects of industrialization and urbanization on agricultural modernization. *Economist* **2016**, *8*, 55–64. [CrossRef]
28. Rosenthal, S.; Strange, W. Geography, industrial organization and agglomeration. *Rev. Econo. Stat.* **2003**, *85*, 377–393. [CrossRef]
29. Zhang, B.; Zhang, J.; Miao, C. Urbanization Level in Chinese Counties: Imbalance Pattern and Driving Force. *Remote Sens.* **2022**, *14*, 2268. [CrossRef]
30. Wang, F.; Qin, Y.; Liu, J.; Wu, C. A study on the influencing factors of carbon emissions from the perspective of multi-dimensional urbanization. *China Popul. Resour. Environ.* **2017**, *27*, 151–161. [CrossRef]
31. Zhao, J.; Liu, Y.; Zhu, Y.; Qin, S.; Wang, Y.; Miao, C. Spatiotemporal differentiation and influencing factors of the coupling and coordinated development of new urbanization and ecological environment in the Yellow River Basin. *Resour. Sci.* **2020**, *42*, 159–171. [CrossRef]
32. Yao, F.; Zhu, H.; Wang, M. The impact of multiple dimensions of urbanization on co2 emissions: A spatial and threshold analysis of panel data on China's prefecture-level cities. *Sustain. Cities Soc.* **2021**, *73*, 103113. [CrossRef]
33. Xu, H.; Jiao, M. City size, industrial structure and urbanization quality—A case study of the Yangtze River delta urban agglomeration in China. *Land Use Policy* **2021**, *111*, 105735. [CrossRef]
34. Liu, W.; Jiao, F.; Ren, L.; Xu, X.; Wang, J.; Wang, X. Coupling coordination relationship between urbanization and atmospheric environment security in Jinan City. *J. Clean. Prod.* **2018**, *204*, 1–11. [CrossRef]
35. FAO. *The State of Food Insecurity in the World 2013: The Multiple Dimensions of Food Security*; FAO: Rome, Italy, 2013.
36. Zhang, Y.; Liu, C.; Guo, L. Appraisal and strategic consideration on food security status of China. *China Rural Surv.* **2015**, *1*, 2–14.
37. Godenau, D.; Caceres-Hernandez, J.J.; Martin-Rodriguez, G.; Gonzalez-Gomez, J.I. A consumption-oriented approach to measuring regional food self-sufficiency. *Food Secur.* **2020**, *12*, 1049–1063. [CrossRef]
38. Ruan, H.; Yu, J. Changes in land over and evapotranspiration in the five Central Asian countries from 1992 to 2015. *Acta Geogr. Sin.* **2019**, *74*, 1292–1304. [CrossRef]
39. Zhang, Y.; Pu, Y.; Chen, L. Urbanization and service industry agglomeration: A view based on system coupling interaction. *China Ind. Econ.* **2013**, *6*, 57–69. [CrossRef]
40. Pan, J.; Chen, Y.; Zhang, Y.; Chen, M.; Fennell, S.; Luan, B.; Wang, F.; Meng, D.; Liu, Y.; Jiao, L.; et al. Spatial-temporal dynamics of grain yield and the potential driving factors at the county level in China. *J. Clean. Prod.* **2020**, *255*, 120312. [CrossRef]
41. Zhang, W.; Wang, M. Spatial-temporal characteristics and determinants of land urbanization quality in China: Evidence from 285 prefecture-level cities. *Sustain. Cities Soc.* **2017**, *38*, 70–79. [CrossRef]
42. Dagum, C. A new approach to the decomposition of the Gini income inequality ratio. *Empir. Econo.* **1997**, *22*, 515–531. [CrossRef]
43. Han, H.; Ding, T.; Nie, L.; Hao, Z. Agricultural eco-efficiency loss under technology heterogeneity given regional differences in China. *J. Clean. Prod.* **2019**, *250*, 119511. [CrossRef]
44. Zhang, H.; Cheng, M.; Luo, J. The spatial pattern, regional differences and distribution dynamics of rural credit cooperatives' double performance. *J. Quant. Tech. Econ.* **2020**, *37*, 175–192. [CrossRef]
45. Gao, K.; Yuan, Y. Spatiotemporal pattern assessment of China's industrial green productivity and its spatial drivers: Evidence from city-level data over 2000–2017. *Appl. Energy* **2022**, *307*, 118248. [CrossRef]
46. Xu, Y.; Zhao, J. Measurement of the coupling degree of innovation elements allocation and economic high-quality development in China. *J. Quant. Tech. Econ.* **2021**, *38*, 46–64. [CrossRef]
47. Ghorbani, M.; Motallebi, M. The study on shadow price of greenhouse gases emission in Iran: Case of dairy farms. *Res. J. Environ. Sci.* **2009**, *3*, 466–475. [CrossRef]
48. Hou, M.; Yao, S. Spatial-temporal evolution and trend prediction of agricultural eco-efficiency in China: 1978-2016. *Acta Geogr. Sin.* **2018**, *73*, 2168–2183. [CrossRef]
49. Sun, Y.; Xia, Q.; Zhou, M. The economic growth effects of new urbanization development and industrial structure change. *J. Quant. Tech. Econ.* **2016**, *33*, 23–40. [CrossRef]
50. Fan, Y.; Lu, Y.; Zhao, J. Multi- scale analysis of regional differences of grain production in China. *Econ. Geogr.* **2014**, *34*, 124–130. [CrossRef]

51. Putra, A.S.; Tong, G.; Pribadi, D.O. Food security challenges in rapidly urbanizing developing countries: Insight from Indonesia. *Sustainability* **2020**, *12*, 9550. [CrossRef]
52. Wenban-Smith, H.; Faße, A.; Grote, U. Food security in Tanzania: The challenge of rapid urbanization. *Food Sec.* **2016**, *8*, 973–984. [CrossRef]
53. Xu, Z.; Song, Z.; Deng, A.; Chen, W.; Chen, F.; Zhang, W. Regional changes of production layout of main grain crops and their actuation factors during 1981-2008 in China. *J. Nanjing Agric. Univ.* **2013**, *36*, 79–86. [CrossRef]
54. Amorim, W.; Deggau, A.B.; Gonalves, G.; Prasath, A.R.; de Andrade Guerra, J.B.S.O. Urban challenges and opportunities to promote sustainable food security through smart cities and the 4th industrial revolution. *Land Use Policy* **2019**, *87*, 104065. [CrossRef]
55. Liu, X.; Xu, Y.; Engel, B.A.; Sun, S.; Zhao, X.; Wu, P.; Wang, Y. The impact of urbanization and aging on food security in developing countries: The view from Northwest China. *J. Clean. Prod.* **2021**, *292*, 126067. [CrossRef]

foods

MDPI

Review

Measuring Food Insecurity: The Problem with Semantics

Rachel M. Sumsion, Haylie M. June and Michael R. Cope *

Department of Sociology, Brigham Young University, Provo, UT 84602, USA; rm553@byu.edu (R.M.S.); hjune@byu.edu (H.M.J.)
* Correspondence: michaelrcope@byu.edu

Abstract: Rising rates of hunger and food insecurity have sparked a major re-evaluation of all aspects of food systems. Because of the multifaceted nature of food insecurity, however, determining what actions should be taken is challenging, especially since reevaluation efforts are led by experts from several disciplines and there is no consensus about which indicators should be used and how they should be measured. Confusion surrounding the meaning of the terms 'food security' and 'food insecurity' has contributed to this lack of consensus. As indicators inform action, such confusion has slowed those committed to alleviating hunger in identifying the most pressing targets. This review highlights (1) the importance of clearly defining food security and food insecurity and (2) how such definitions affect measures of food insecurity in the United States. While some might say that definitions are an issue of the past or a trivial matter of semantics, we believe that the world's present rates of hunger and malnutrition are attributable, at least in part, to the lack of consensus on these definitions and their accompanying measurements and indicators. Although the present review can be helpful to academics and policy makers, the primary purpose is to be a resource to those involved in the day-to-day production of food, such as ranchers and farmers by providing an overview of definitions, indicators, and measurements used when discussing food security.

Keywords: food insecurity; food security; measurement; hunger; levels of analysis

1. Introduction

The right to food has been recognized as an international human right for almost 75 years [1]. Since the 1950s, dozens of organizations have worked to alleviate hunger and improve food systems and access to food [2]. However, despite these efforts, progress in combatting hunger and food insecurity has stalled and, in some cases, even been reversed in recent years due in part to the COVID-19 pandemic and Russian–Ukrainian conflict. Indeed, "after remaining relatively unchanged since 2015 ... the number of people affected by hunger since (2019 grew) by about 150 million" [3]. In 2020, an estimated 828 million people went hungry [3], prompting researchers and policymakers to re-evaluate global food systems [4]. Unfortunately, progress in such efforts has been slowed, however, due to various understandings of the terms 'food security' and 'food insecurity' as well as their accompanying indicators.

Research on food insecurity spans a variety of disciplines. Agriculturists, anthropologists, economists, nutritionists, sociologists, and others have all made valuable contributions on the subject [5]. The value of these multidisciplinary efforts cannot be understated; however, significant variation exists between each discipline's definitions and methods of measuring food security. Despite international consensus supposedly being reached in 1996 on the definition of food security (the 1996 World Food Summit definition of food security (updated in 2009) is: "Food security exists when all people, at all times, have physical, social and economic access to sufficient, safe and nutritious food to meet their dietary needs and food preferences for an active and healthy life" [2]), many definitions of 'food security' and 'food insecurity' are still commonly used and propagated today between and within disciplines. As definitions precede measurements, measurements produce indicators, and indicators inform action, this is problematic [6]. Comprehensive, valid measurements of food security are needed to ensure accuracy in current and future projections of food insecurity. Such measurements are built upon a foundation of common definitions. This review serves as a call to reestablish consensus in the definitions and measures of food security and insecurity as without agreement on who is experiencing food insecurity, where they are located, when they are most vulnerable, or why they are experiencing it, little can be done to effectively combat it [6].

This paper presents an overview of food security/insecurity, by summarizing current and common definitions of the terms, as well as different indicators and measurements of said indicators. It highlights notable similarities and differences between definitions and makes recommendations for a renewed focus to be placed on the use of consistent definitions between and within disciplines. Following this semantic overview, we illustrate how different definitions of food security and insecurity affect measurements and indicators of the topic in the United States. To conclude, we recommend that standardized definitions are reviewed and reemphasized.

Although some experts may say that definitions are an issue of the past or that this topic should be left to semanticists, we believe that the world's tragic rates of hunger and malnutrition are partially attributable to the confusion surrounding what 'food insecurity' is and how to measure it [7]. The current paper contributes to the literature by providing a comprehensive review of the language of 'food insecurity' [8]. While the current paper can be useful to academics across disciplines and to policy makers, the primary audience of the paper are members of communities that contribute to food production on a day-to-day basis, such as farmers and ranchers. We believe that providing a comprehensive review for this audience can be used to inform policy, help those in food production receive needed government grants, and ultimately help alleviate food insecurity.

2. Review

2.1. Why Food Insecurity?

Since 1948, the right to food has been internationally recognized as a basic human right [9]. Despite a broad consensus about what this means, together with many valiant efforts to combat malnutrition in all its forms, it is estimated that over 820 million people worldwide go hungry and around 2 billion people experience food insecurity each year [4,10,11]. Food insecurity is a serious issue as it is related to a variety of adverse physical, mental, and social health outcomes for both adults and children. Adults who experience food insecurity are more likely to contract noncommunicable diseases (such as diabetes, hypertension, hyperlipidemia, or chronic heart disease), have a reduced fertility rate, a decreased basal metabolic rate, micronutrient deficiencies (particularly protein, Vitamin A, B Vitamins, calcium, iodine, iron, and zinc), food allergies, or be overweight or obese [4,9,12–17]. Children experiencing food insecurity have higher rates of cognitive, behavioral, and social problems, and are more likely to be stunted, wasted, have poor oral health, anemia, asthma, or suicidal ideations [8,13,15,18–20]. These adverse effects, along with others, have been linked to people experiencing food insecurity spending over USD 1860 more per capita annually on healthcare than people who are food secure [15]. In the most extreme cases, food insecurity can lead to death for both children and adults.

In response to these grim findings, and in recognition that food security is one of the world's most pressing issues, government and food security experts have been meeting for years to discuss hunger and food security [10,21]. Their combined efforts have helped to reduce the number of those affected by hunger and malnutrition. In 2016, the United Nations (UN) announced that its second Sustainable Development Goal was to "end hunger, achieve food security and improve nutrition and promote sustainable agriculture" by 2030. To do so, they planned to double small farmers' productivity and incomes, increase agriculture investment, and ensure food markets' proper functioning [15,22]. While the UN's inspiring goal invigorated efforts to improve food systems, and a significantly larger number of studies on food insecurity—including issues of sustainability, the environment, socioeconomics, culture, politics, and governance—have been conducted over the past nine years, rates of worldwide food insecurity have been on the rise since 2014 [4] with recent reports projecting that nearly 670 million people will still be undernourished in 2030 [3].

2.2. Defining Food (In)Security

'Hunger' is a familiar, but multi-layered term [23]. In academia, scholars often refer to it as the "painful or uneasy sensation caused by a lack of food" [24] (p. 1560) [25]. However, due to its subjective nature, since 2006 there has been no officially recognized metric of 'hunger' [26]. Instead, researchers and policymakers measure the food security and food insecurity of nations, regions, households, and individuals—levels of analysis which seldom yield the same results. While 'hunger' and 'food insecurity' are not synonymous, they are closely related, with hunger being a potential consequence of food insecurity [27,28].

Ironically, while measures of 'hunger' have been rejected for their subjectivity, the concepts of 'food security/insecurity' are also defined and measured without objectivity [29]. While the terms have been around since the 1950s, they are understood and applied differently according to the context and region in which they occur [30,31]. The term 'food security' was first defined at the World Food Conference in 1974 as "(the) availability at all times of adequate world food supplies of basic foodstuffs to sustain a steady expansion of food consumption and to offset fluctuations in production and prices" [32]. As understanding of the causes and consequences of food security/insecurity have shifted and expanded since that time, however, definitions have continuously evolved. Food security/insecurity has been described as a flexible, multidimensional concept, and has been understood and applied differently by sociologists, agriculturists, political scientists, nutritionists, and economists throughout the years [33,34]. In the past especially, various viewpoints contributed to a "bewildering number of paradigms and points of view" in the literature [9] (p. 50). For instance, in the 1990s, Smith, Pointing, and Maxwell said that "there is no single definition (of food security or food insecurity) ... but rather a complex weave of inter-related strands which are adjusted to suit the needs and priorities of individual users" [35] (p. 136). To illustrate their point, they compiled an annotated bibliography of close to two hundred definitions of these terms and called for the development of a comprehensive definition [35]. As a result, at the 1996 World Food Summit, a new definition of food security was proposed and ratified [36], to which only minor adjustments have been made since. The most recent agreed upon definition states that:

Food security exists when all people, at all times, have physical, social, and economic access to sufficient, safe, and nutritious food that meets their dietary needs and food preferences for an active and healthy life [36].

While this definition has largely been accepted and is used by most scholars well-versed in the literature, definitions of food security and its counterpart, food insecurity, still vary between disciplines and countries. For instance, while the Life Sciences Research Office's 1990 definition of food insecurity (food insecurity is the limited or uncertain availability of nutritionally adequate and safe foods, or the limited or uncertain ability to acquire acceptable foods in socially acceptable ways [24]) is often used in the United States, it has not been widely adopted elsewhere.

While Smith et al.'s work helped clarify the concepts of food security and food insecurity in the 1990s, a good number of definitions of the concepts are still used and propagated today in academic as well as grey literature. Inspired by and drawing from the work of Smith et al., Tables A1 and A2 illustrate the relevance of this issue as each of the 52 definitions presented (19 for food security and 33 for food insecurity) have been cited as primary definitions of the concepts in scholarly literature since 2000 (four years after the UN's World Food Summit definition was ratified to allow time for adoption). While we acknowledge that the definitions of individual researchers unfamiliar with the semantic history of food insecurity should not be given the same weight as definitions produced through more intensive processes, we maintain that to non-academic audiences and/or those less familiar with food security literature, the continued use of so many definitions can have negative repercussions, especially since much of the nuance between these definitions is far from insignificant when it comes to measuring food insecurity [9,10,35,37].

2.3. Varying Definitions of Food Security

If poor data leads to poor policies, then rising rates of food insecurity may be attributable, at least in part, to a lack of consensus in the measures and corresponding indicators of 'food security' and 'food insecurity.' Definitions, targets, and cut-off points for indicators such as 'adequate', 'sufficient', 'nutritious', 'acceptable', or 'preferred' foods vary from organization-to-organization, fragmenting the efforts of organizations committed to eradicate hunger, despite reaching a consensus during the World Food Summit in 1996 [7]. The World Food Summit definition has varying dimensions and while it is widely adopted in the literature in the United States, consistent indicators and measurements have not. To better inform policy decisions and develop more effective interventions, clear data, indicators, and measurements are needed. Each of these aspects is dependent upon sound definitions and clarity in regard to which specific dimensions of a definition are being measured [38,39].

Without concretely defining the bounty of food security or the threat of food insecurity, actionable solutions to the latter may remain out of reach [40]. We recognize that there is danger in relying on a single definition as a "definition implies a choice, a particular way of seeing a problem among a range of alternatives (and) policy is determined in part by that choice" [41] (p. xii). Accordingly, some scholars plausibly argue that a single, universal, "catch-all" definition does not adequately address the complexities of food security or food insecurity [5,37,39,42,43] (p. 1023–1024). For this reason, multifaceted definitions are needed to generate broad agreement among policymakers and practitioners with diverse emphases. Without such agreement, inconsistencies will continue to exist between various organizations' measurements, policies, and actions [16]. Care should be taken not to conflate 'food security' or 'food insecurity' with other terms. For instance, in many cases, a 'food security' organization's focus may be more appropriately described by terms such as 'nutrition security,' 'hidden hunger,' 'food insufficiency,' 'food access,' 'food capacity,' 'food resilience,' 'food rights,' or 'food sovereignty' [5,44,45]. Additionally, the multi-dimensionality of food security should be recognized when developing and using measurements for it [39].

For effective policies to be made, common measurements and agreed upon definitions of 'food security' and 'food insecurity' are needed [8,38]. Tables A1 and A2 present many definitions that have been used to describe these concepts. Most broadly accepted definitions recognize that 'food security' and 'food insecurity' are multifaceted concepts. 'Four pillars' are commonly agreed upon [8,46,47]: food availability (supply and production), food access (economic and physical), food utilization (use), and food stability (consistency) [15]. In recent years, calls to expand our conceptual definition of food security has led to the proposal of two additional pillars: agency (empowerment in consumption and production) and sustainability (long-term impacts) [37,48]. Despite such similarities, differences in aspects such as the level of analysis to which the definition applies, what is meant by phrases such as 'enough food,' 'adequate food,' 'preferred food,' or 'culturally or socially acceptable means,' and the relative importance or unimportance of self-perceptions impact measurements, rates, and understandings of the severity of food security and insecurity. Such differences are crucial for policymakers to be aware of and understand. We address this point in the following section by exploring how definitional differences can lead to variation in measures of food security/insecurity by implying different levels of analysis and understandings of phrases.

2.3.1. Levels of Analysis

Although hunger is a state that is experienced on the individual level, food security and food insecurity are often conceptualized at various levels of analysis [18,35]. The most common levels of analysis are 'global,' 'national,' or 'regional' (which generally concern the level and reliability of the availability of food supplies), 'household' (which generally means the household's economic, physical, or socio-cultural access to food), or 'individual' (often someone's economic, physical, or socio-cultural access and entitlement to food) [31,49]. Table 1 summarizes each of the levels of analysis.

Table 1. Levels of analysis.

Level of Analysis	Description
Global	Refers to the availability and reliability of food worldwide.
National	Refers to the availability and reliability of food at a national level, the production of food within countries, and the levels of food reserves that should be maintained consistently.
Regional	Refers to the availability and reliability of food at a regional level.
Household	Refers to a household's physical and economic access to food, their levels of vulnerability, and their utilization of food.
Individual	Refers to an individual's physical and economic access to food (recognizing that food is not always evenly distributed at the household level), their levels of vulnerability, and their utilization of food.

Note: The table gives descriptions of different levels in which experiencing hunger is measured.

Before the 1980s, food insecurity was considered almost exclusively at the global or national level and was understood to be an issue of food supply; that is, the main driver of food insecurity was thought to be the lack of food [48]. Consequently, most efforts to address food insecurity focused on improving agricultural yields and regulating national reserves. Over time, however, researchers recognized that the availability of food alone was not sufficient to combat hunger. This fact was widely accepted in the early 1980s with the publication of Amartya Sen's Poverty and Famines in which he argued that a lack of food entitlement—unequal access to food by production, trade, one's own labor, or transfers—not a lack of food itself, was a principal driver of food insecurity [50]. From this time forward, "food access" which is measured at the household and individual level has been recognized as a principal component of food security [9,50].

While progress has accompanied this analytical shift, choosing to operationalize food security/insecurity at any given level of analysis requires making assumptions about that level [9]. Consider the 'household' level of analysis. What is meant by 'household'? While it was originally intended to mean individual family units, today some consider any occupants under a single roof, including non-family members such as tenants, boarders, roommates, or other non-dependents as 'household' members.

Further, assuring that enough food is allocated at the household level does not necessarily translate to people receiving enough food at the individual level as the distribution of food within households can still be unequal to people's specific needs [8,50]. For instance, maternal buffering occurs when mothers consume less food to give their children more [1]. Such buffering, along with other similar coping strategies, would pass undetected if organizations solely relied upon household level measures. If organizations relied solely upon individual level measures, however, they could fall victim to overestimating the severity of such a family's experience. Suffice it to say, in cases when levels of analysis are conflated, rates of food insecurity can be over-or-under-exaggerated [50]. For this reason, deciding what level of analysis the concepts of food security and insecurity are meant to refer to might clarify the mission of organizations and effectively narrow their focus in helping those experiencing food insecurity. Of course, the goal of doing so would not be to ignore other levels of food security or insecurity, rather to clarify those that are most important to organizations. Terms such as 'food supply' (global, national, or regional levels of analysis) or 'food access' (household or individual levels of analysis) might be profitably used to clarify these differences.

2.3.2. Same Words, Different Ideas

Cultural differences and expectations impact the interpretations of definitional constructs, or the different dimensions of a definition. In measuring food security and insecurity, these different understandings can have a significant impact on food security data. For instance, what is meant by 'enough' or 'sufficient levels' of food significantly affects reported rates of food insecurity. 'Enough' can refer to a variety of indicators such as individual energy requirements (measured as calories or macronutrients (often revised according to age, gender, and activity rates) [8] (p. 1), personal nutrition requirements (measured in micronutrients; specifically, vitamin A, iron, and iodine [9]), or personal perceptions (measured according to self-reported surveys [50]). Several of the variations in how such indicators are used can be explained by the definitions influencing them. Some definitions imply that 'enough' means a minimal level of food consumption (see 51–54), others mean the amount of food required to meet nutritional needs (see 18, 24, 55–59), and still others mean the amount necessary to live an "active, healthy life" (see 2, 26, 24, 50, 60–67) (see Table 2).

Table 2. Different Interpretations of 'Enough'.

Citation	Definition
	Minimal levels of food consumption
Alamgir and Arora 1991 [51]	The sum of household and sub-national food security, and more. At the national level, food security can be defined as assured national availability of food to meet current *minimum requirements per capita* during a reference period (normally, one year) and, also, to meet any unexpected shortfall over a limited period (e.g., three months).
IFAD, 1991 [52]	Access to enough food to ensure the *minimum necessary food intake* for all individual members to lead a healthy life.
Siamwalla and Valdes 1980 [53]	The ability to meet *target levels of consumption* on a yearly basis.
World Bank Staff, 1980 [54]	The assurance of a *minimally adequate level of food consumption*.
	Necessary to meet nutritional needs
Alaimo, 2005 [18]	Food insecurity: limited or uncertain availability of *nutritionally adequate* or safe foods.
Barraclough and Utting 1987 [55]	Assured access by all social groups and individuals to food *adequate in quality and quantity to meet nutritional needs*.
Benson et al., 1986 [56]	Having assured sets of entitlements—from food production, cash income, reserves of food or assets, and/or government assistance programs—such that in times of need people will be able to maintain *sufficient nutritional intake for physical well-being*.
Eicher and Staatz, 1986 [57]	The ability of a country or region to assure, on a long-term basis, that its food system provides the total population access to a timely, reliable, and *nutritionally adequate supply of food*.
Jonsson and Toole, 1991 [58]	Access to food, *adequate in quantity and quality, to fulfill all nutritional requirements* for all household members throughout the year.
Life Sciences Research Office and Andersen, 1990 [24]	Food insecurity: the limited or uncertain availability of *nutritionally adequate* and safe foods or limited or uncertain ability to acquire acceptable foods in socially acceptable ways.
Winne et al., 2000 [59]	All persons in a community have access to culturally acceptable, *nutritionally adequate* food through local non-emergency sources at all times.
	The amount of food needed for an "active, healthy life"
Bartfeld and Dunifon, 2006 [60]	The assured access of all people to *enough food for a healthy and active life*.
Coleman-Jensen et al., 2021 [61]	Consistent, dependable access to *enough food for active, healthy living*.
FAO et al., 2009 [2]	Food security exists when all people, at all times, have physical, social, and economic access to sufficient, safe, and nutritious food that meets their dietary needs and food preferences *for an active and healthy life*.
Haddad et al., 1995 [62]	Availability of sufficient food at all times for all people to *ensure an active and healthy life*.
Hayes, 2021 [26]	Consistent access to *enough food for active, healthy lives* for all household members at all times during the year.
Kabeer, 1990 [63]	The ability of a household to assure all its members sustained access to *sufficient quantity and quality of food to live active, healthy lives*.
Life Sciences Research Office and Andersen, 1990 [24]	Access by all members at all times to *enough food for an active, healthy life*. Food security includes at a minimum:the ready availability of nutritionally adequate and safe foods; assured ability to acquire acceptable foods in socially acceptable ways (i.e., without resorting to emergency food supplies, scavenging, stealing, or other coping strategies).
Pinstrup-Andersen, 2009 [50]	Access by all people to *enough food to live a healthy and productive life*.
Reutlinger, 1985 [64]	Access by all people at all times to *enough food for an active and healthy life*.
Reutlinger, 1986 [65]	Access by all people at all times to *enough food for an active, healthy life*.
United Nations, 1990 [66]	The ability of household members to assure themselves sustained access to a *sufficient quantity and quality of food to live active, healthy lives*.
World Food Program, 1996 (2009) [67]	When all people, at all times, have physical and economic access to sufficient, safe and nutritious food that meets their dietary needs and food preferences *for an active and healthy life*.

Note: The table gives definitions and citations of different interpretations of 'enough' in the literature.

Similar analyses of definitional constructs such as 'adequate foods', 'preferred foods', and 'culturally or socially acceptable means' are also important to understand (see Tables A3–A5). Although some scholars have made efforts to clarify these constructs (for instance, Pinstrup-Andersen defines 'preferred foods' as "foods that are socially and culturally acceptable and consistent with religious and ethical values" [50] and Frongillo and Horan define 'socially unacceptable means' as "buying food on credit, using food pantries, or ... asking friends and relatives for money for food" [47] (p. 30), most definitions lack such specification. We believe that each of these constructs needs clarifying to be of value to policymakers. In their present form, they are largely subjective, and their applications vary significantly between contexts [39]. This presents difficulties in developing comparable measurements and coordinated plans for action. By finding common ground for definitions and measurements, a more comprehensive understanding of food security and food insecurity will emerge and lead to meaningful progress in combatting world hunger.

2.4. Measuring Food Security

Measures of food security are continually being refined and improved, even though scholars generally agree than food security rests on the four pillars of availability, access, utilization, and stability [37,48,68]. In 2006, Webb et al. identified that many measurements were (1) shifting their focus from measures of availability and utilization to measures of access, (2) shifting from objective to subjective measures, and (3) placing more of an emphasis on direct rather than proxy measures of food security [69]. In 2012, the Committee on World Food Security recommended that all measurements of food insecurity be based on human rights, ensure accountability, involve all relevant stakeholders, be easily understandable, and build upon rather than duplicating national capacities [68]. Recently, some food security definitions have expanded to include agency and sustainability as pillars to understanding the right to food [37,48].

Some of the most widely recognized measures of food security in the United States today include Anthropometric measures, Coping Strategies Index (CSI), Domestic Food Price Volatility, Food Consumption Score (FCS), Food Insecurity Experience Scale (FIES), Global Food Security Index (GFSI), Global Hunger Index (GHI), Household Consumption and Expenditure Surveys (HCES), Household Dietary Diversity Score (HDDS), Household Food Insecurity Access Scale (HFIAS), Household Pulse Survey (HPS), Months of Adequate Food Provisioning (MAHFP), Prevalence of Undernourished (PoU), and the Relative Dietary Supply Index (see Table 3). Each of these indicators assess slightly different aspects of food security. For instance, while some of these indicators focus on macro-measures at the global, national, or regional levels of analysis, others focus on micro-measures at the household and individual levels. Consequently, different measures prioritize selective pillars of food security as food supply (availability) is typically measured on a macro-scale while economic or physical access, dietary diversity (utilization), and coping strategies (stability) are typically measured on micro scales [5,70]. Every indicator yields important insights, but sizeable gaps remain between the multifaceted nature of food security and what existing measurements reveal about the concept [39,41]. As the concept of food security has begun to be expanded to include sustainability and agency, questions about measurements of these indicators have been brought up [37].

Table 3. Measures of food security in the United States (2022).

Measure	Description
Anthropometric measures	Measures bodily characteristics such as height, weight, skinfold, etc. (combined with weight and age) to assess the utilization of food [5]
Coping Strategies Index (CSI)	Measures the frequency and severity of context specific behaviors taken in response to food insecurity [5]
Domestic Food Price Volatility	Measures the variability in the annual food price index [5]
Food Consumption Score (FCS)	Measures the quality, quantity, and diversity of foods consumed by a household in the last seven days [70]
Food Insecurity Experience Scale (FIES)	Measures the severity of food insecurity in population groups based off individual or household level data [68]
Global Food Security Index (GFSI)	Measures the affordability, availability, quality, resiliency, and safety of available foods in given nations [15]
Global Hunger Index (GHI)	Measures rates of undernourishment, underweight children, and child mortality at the national level to assess hunger [68,70]
Household Consumption and Expenditure Surveys (HCES)	Measures a variety of household socioeconomic conditions affecting food security such as food acquisition and consumption [5]
Household Dietary Diversity Scores (HDDS)	Measures the number of food groups (grains, starches, vegetables, fruits, meat, eggs, fish, legumes, dairy, fats, sugar, condiments, etc.) consumed by an individual or household in the previous 24 h [71]
Household Food Insecurity Access Scale (HFIAS)	Measures the frequency and intensity of challenges that households face in accessing food [71]
Household Food Security Survey Module (HFSSM)	Measures the prevalence of food insecurity among households in the United States [49]
Household Pulse Survey (HPS)	Measures household sufficiency of food in the last 7 days, 2 weeks, or month based off a single question [26,28]
Months of Adequate Food Provisioning (MAHFP)	Measures a household's stability in maintaining adequate levels of food in the past year [5,71]
Prevalence of Undernourished (PoU)	Measures the percentage of a country's population whose typical food intake is below the levels needed for an active and healthy life using national estimates [68,72]
Relative Dietary Supply Index	Compares the available dietary energy supply in a country with that country's average caloric needs [5]
Sustainable Nutrition Security (SNS)	Incorporates metrics of sustainability [73]
Women's Empowerment Nutrition Grid (WEN or WENI)	Incorporate agency, alongside knowledge and resources, as dimensions of empowerment with respect to food security [37,74,75]

Note: The table gives measures and their descriptions of food security in the United States.

As no single measure is adequately represents all aspects of food insecurity, many organizations have begun triangulating their use of measurements and indicators [68,71]. Some have found value in utilizing indirect measurements such as food balance sheets, temperature, rainfall, marketing data, political instability, trade policies, or household size and dependency ratios to capture the concept more fully [1,5,39]. Unfortunately, some have erroneously used measurements interchangeably, failing to recognize that not all methods capture the same thing [5,39]. To tackle this issue, some have recommended that interchangeable measures of food security be developed [71].

2.5. Common Measures of Food Insecurity

In the ensuing sections, we present a few of the most common measurements of food security in the United States with their associated advantages and limitations to highlight how difference in measurements may impact our understanding of food security.

2.5.1. US Household Food Security Survey Module and Food Insecurity Experience Scale

The US Household Food Security Survey Module (HFSSM) and Food Insecurity Experience Scale (FIES) are two of the most used measures of food insecurity in the United States. They are similar experience-based metrics that directly ask people about their access to adequate food in a given reference period [68]. Both measures use yes-no questions to determine individual's level of food insecurity, starting with the least severe consequences of food insecurity (worry about acquiring food) and progressively moving towards more severe consequences (hunger). To illustrate, the first question of the FIES asks "In the last year was there a time that you were worried you would not have enough food to eat because of a lack of money or other resources?", the third question asks, "In the last year was there a time when you only ate a few kinds of foods because of a lack of money or other resources?" the fifth question asks "In the last year was there a time when you ate less than you thought you should because of a lack of money or other resources?", and the eighth and final question of the FIES asks, "In the last year was there a time when you went a whole day without eating because of a lack of money or other resources?" [76]. The HFSSM follows a very similar structure, but for households with children, it asks an additional ten questions that specifically relate to children's access to food [28]. Using data collected by these measures, it is possible to estimate how many people in a given population are experiencing food insecurity, as well as how extreme their experience with food insecurity may be [68,77,78].

While similar in many ways, the HFSSM and the FIES do have some differences. Of the two, the HFSSM (also known as the Core Food Security Module (CFSM)) was developed first by the US Department of Agriculture through the Community Childhood Hunger Identification Project in 1995 [49,68]. Since that time, the HFSSM has been widely used to research the causes and consequences of food insecurity at the household level in the United States and is administered today as part of the monthly Current Population Survey (CPS) and the bi-annual National Health and Nutrition Examination (NHANES) [20,69]. As was mentioned, the HFSSM can be adapted for families with (18 questions) and without children (8 questions). Its reference period can also be adapted, although typically it is used to gather data about people's experience with food insecurity in the past year or in the past 30 days [49]. It (along with a shortened 6-item version) has been validated in effectively measuring people's psychological experience with and economic access to food but falls short in adequately capturing other dimensions of food access and other pillars of food security [38,47,79]. The HFSSM is also lengthy and cannot be used to measure food security on an individual level; therefore, it fails to provide an effective cross-national measure of food insecurity.

The FIES was developed by the Food and Agriculture Organization (FAO) through the Voices of the Hungry Project in 2014 [80]. Its similarity to the HFSSM can be attributed to its basis in both the HFSSM and Latin America and Caribbean Food Security Scale (ELCSA) [68]. In fact, the FIES was specifically developed to address some of the shortfalls of the HFSSM. Accordingly, the FIES is a shorter, more standardized experience-based measure of food insecurity that can be adapted to assess food insecurity at either the individual or household level. Going one step further than the HFSSM, it can be used to measure the percentage of individuals in the population who have experienced food insecurity at moderate or severe levels during a given reference period [80,81]. Initially administered in the Gallup Poll, the FIES has been recognized as a valuable global measure and has been translated into 170 languages and dialects. It has been culturally attuned and provides a standardized measurement of food security across countries despite vast cultural and linguistic differences [76,80]. As the FIES has become increasingly accepted worldwide, it has been used by the FAO (in conjunction with the PoU) to measure the world's progress toward its Sustainable Development Goals [68,80]. Notable strengths and weaknesses of the FIES are listed below (see Table 4).

Table 4. Strengths and weaknesses of the FIES.

Strengths	Weaknesses
Helps identify risk factors and consequences of food insecurity. Can assist in identifying vulnerable populations before times of crisis. Effectively captures psychosocial aspects of food security. Simple, time effective, and relatively inexpensive Provides a standardized and comparable global measurement. Allows disaggregation of data by gender. Can be used in combination with other indicators. Has been effectively translated into many languages.	Does not capture the full range of food security. Does not measure diet quality, food consumption, or expenditures. May be challenging for non-specialists to analyze data. Assumes that the process of food insecurity is orderly and predictable across all cultures. Is relatively new.

Note: Author's compilation. Source: [68,72,80].

2.5.2. Household Food Insecurity Access Scale and Household Hunger Scale

Another common measure is the Household Food Insecurity Access Scale (HFIAS). The HFIAS was developed by the United States Agency for International Development (USAID) and is like the HFSSM and FEIS in that it is primarily used to measure a household's access to food [68,71]. It has a 30-day recall period and includes nine questions about psychological and behavioral factors influencing food security at the household level [49,70]. These questions ask about people's feelings of uncertainty or anxiety related to food supply, as well as the preferability, variety, and quantity of foods consumed at the household level [82]. While the HFIAS was initially developed for program monitoring and impact evaluation, it has been used in other circumstances as well [49]. Critics of the HFIAS have pointed out that the reports gathered through this method are subjective and not necessarily applicable across cultures [5].

In response to these critiques, an improved version of the HFIAS, the Household Hunger Scale (HHS), was recently developed. The HHS is very similar to the HFIAS, but only has three questions that measure the most severe aspects of food insecurity [5]. As more severe experiences of food insecurity are less subjective (e.g., running out of food is less subjective than feeling anxious about running out of food), the HHS is believed to provide more comparable results across countries and contexts [49]. Despite this improvement, both the HFIAS and HHS have additional weaknesses, the most notable being that they both fail to adequately measure other pillars of food security such as availability, utilization, or stability.

2.5.3. Prevalence of Undernourishment

Food availability is often assessed using the Prevalence of Undernourishment (PoU) indicator [49]. The PoU measure utilizes national food supply, consumption, and energy needs data to estimate the percentage of a country's population whose typical food intake is below minimal consumption levels [5,68,72]. The FAO uses this indicator (in conjunction with the FIES) to estimate the number of people who are likely not eating enough food to meet their dietary energy needs at the national level [68,78]. This indicator is useful in identifying countries in need and in making comparisons at the national level, but it does not adequately identify vulnerable populations as it provides no specific data on what households or individuals are food insecure or where those experiencing it live [5,29,68,78].

2.5.4. Coping Strategy Index

The Coping Strategy Index (CSI) was developed by the World Food Program and the Cooperative for Assistance and Relief Everyone International (CARE International) and is used to assess the frequency and severity of coping behaviors taken in response to the experience of food insecurity [5,39,70]. It is used by a variety of organizations to help identify at-risk and food insecure households [49,70]. No "universal CSI" exists as every nation, region, and locality has different cultural norms and coping strategies that are engaged in when responding to the experience of food insecurity [5,49]. Accordingly, various culturally attuned versions of the CSI have been developed and are used in different areas of the world [5]. For instance, in more developed nations, a Reduced Coping Strategy Index (rCSI) is often used. This rCSI only measures a couple common, but less-severe coping behaviors, in contrast to the CSI which includes measures of more severe coping behaviors [70]. Because of the variations that exist between CSI measurement tools, it is only to be used as a comparative indicator within a specified geographic area [5]. While the CSI helps policymakers understand people's behavioral responses to food insecurity in a given region, it does not fully capture other important aspects of food security [49].

2.5.5. Household Dietary Diversity Score and Food Consumption Score

The Household Dietary Diversity Score (HDDS) and Food Consumption Score (FCS) are used to assess food utilization. Both measures analyze the quantity and diversity of foods consumed within a given reference period [70]. The HDDS was developed as part of the Food and Nutrition Technical Assistance Project and assesses the number of food groups that a household has consumed in the last 24 h [5]. It can be adapted to specifically capture the experience of individuals (IDDS) or women (WDDS) and is predominately used by the FAO and USAID [70]. Notably, the HDDS has no predetermined cut-off point in establishing an adequate level of dietary diversity.

Like the HDDS, the FCS also assesses the number of food groups that a household has consumed, but typically has a seven-day reference period. The FCS also measures the quantity of foods consumed, applying weights to different food groups relative to their nutritional value [70]. The FCS does have a predetermined cut off point in determining who is or is not consuming adequate levels of dietary diversity, but these cutoffs are relatively arbitrary [49]. The FCS is most often distributed by the World Food Programme [70]. While extremely valuable, the FCS only captures dietary diversity on the household level and does not account for differences in the intrahousehold consumption of nutrients [70].

3. Limitations

Measurement Limitations

Despite the progress that has been made in recent years in measuring food security, significant limitations still exist in assessing these concepts. Most measures require significant time, resources, and technological expertise to be understood and analyzed [68]. In addition to the unavoidable complications accompanying any study involving human participants, a researcher's insufficient understanding of an indicator or combination of indicators may result in a blindly inaccurate representation of food security. Even with a perfect understanding of indicators, however, measurement specific shortcomings complicate the process of determining people's food security status [5,83].

Many conventional efforts used to gather data using these indicators overlook the experiences of 'essential workers,' homeless individuals, migrants, refugees, or others living in marginal housing or on reserves [4,12,20,29,61,84]. This under coverage bias is concerning as these groups are often more likely to report experiencing food insecurity [18]. The length of many food security measurements can also affect the representativeness of samples. While answering the HFSSM's 18 questions has been found to only take an average of four minutes, being presented with such a lengthy questionnaire can still be overwhelming [34]. Partially because of this, in times of recent crisis, the USDA has chosen to rely upon a one-item food sufficiency question in the Household Pulse Survey (HPS) to roughly estimate household's access to food [28,85].

4. Conclusions

Although valiant efforts are underway to combat world hunger and reduce the number of malnourished people, further progress will depend on clarifying the semantics of food security and food insecurity. Despite the ratification of universal definitions in the 1990s, which were meant to curb the development of new definitions, the terms have continued to diverge due to the multifaceted nature of the concepts [86,87]. While most definitions of food security/insecurity used today highlight similar points, slight, yet impactful, semantic differences remain, hindering researchers, practitioners, and policymakers from effectively measuring and finding solutions to food insecurity. By recognizing such semantic differences and the corresponding challenges of existing measurements of food insecurity, policy makers and practitioners can create effective policies that will reverse the rising rates of food insecurity the world is seeing today. Without re-emphasizing the use of consistent definitions and measures of food insecurity, effective actions and policies to combat hunger will remain beyond our reach.

Author Contributions: Conceptualization, R.M.S. and M.R.C.; methodology, R.M.S. and M.R.C.; validation, M.R.C.; formal analysis, R.M.S.; investigation, R.M.S.; resources, M.R.C.; data curation, R.M.S.; writing—original draft preparation, R.M.S.; writing—review and editing, H.M.J. and M.R.C.; visualization, M.R.C.; supervision, M.R.C.; project administration, M.R.C. All authors have read and agreed to the published version of the manuscript.

Funding: This research received no external funding.

Data Availability Statement: No new data were created or analyzed in this study. Data sharing is not applicable to this article.

Conflicts of Interest: The authors declare no conflict of interest.

Appendix

The Tables A1 and A2 provide definitions of food security present in the literature since 1996 and 2000, respectively. The tables provide the citation, level of analysis (individual, national, regional, etc.), and definition.

The Tables A3–A5 provide definitions and their citation of different ideas of 'adequate food', 'preferred foods' and 'culturally or socially acceptable means' based on the literature, respectively.

Table A1. Definitions of Food Security Used in the Literature Since 1996.

Citation	Level of Analysis	Definition
Bartfeld and Dunifon, 2006, p. 921 [60]	Individual	Assured access of all people to enough food for a healthy and active life.
Béné, 2020 [88]	Individual, Household	When individuals and households have adequate resources to obtain appropriate food
Blumberg et al., 1999, p. 1231 [79]	Individual	Assured access to nutritionally adequate and safe foods without resorting to emergency food supplies, scavenging, stealing, and other coping strategies.
Coleman-Jensen et al., 2021 [61]	Individual	Consistent, dependable access to enough food for active, healthy living.
Committee on World Food Security, 2012 [46]	Individual	When all people at all times have physical, social, and economic access to food, which is safe and consumed in sufficient quantity and quality to meet their dietary needs and food preferences, and is supported by an environment of adequate sanitation, health services and care, allowing for a healthy and active life.
Wood et al., 2000 [89]	Individual	The state in which all persons obtain a nutritionally adequate, culturally acceptable diet at all times through local, non-emergency sources.
FAO et al., 2009 [2]	Individual	Food security exists when all people, at all times, have physical, social, and economic access to sufficient, safe, and nutritious food that meets their dietary needs and food preferences for an active and healthy life.
Hayes, 2021 [26]	Individual	Consistent access to enough food for active, healthy lives for all household members at all times during the year.

Table A1. *Cont.*

Citation	Level of Analysis	Definition
Life Sciences Research Office and Andersen, 1990 [24]	Individual	Access by all members at all times to enough food for an active, healthy life. Food security includes at a minimum: the ready availability of nutritionally adequate and safe foods; assured ability to acquire acceptable foods in socially acceptable ways (i.e., without resorting to emergency food supplies, scavenging, stealing, or other coping strategies).
Pinstrup-Andersen, 2009 [50]	Global, National, Regional, Household	Enough food is available, whether at the global, national, community, or household level.
Pinstrup-Andersen, 2009 [50]	Individual	Access by all people to enough food to live a healthy and productive life.
Prifti, 2021, p. 238 [72]	National, Regional, Household, Individual	A function of availability of adequate food in terms of quantity and quality and the people's ability to afford it at all times.
Siche, 2020 [90]	Individual	Everyone has unrestricted access to food that allows them to satisfy their basic needs.
Sustainable Development Commission, 2009, p. 10 [91]	National	The aspiration for genuinely sustainable food systems, where the core goal is to feed everyone sustainably, equitably, and healthily; which addresses needs for availability, affordability, and accessibility; which is diverse, ecologically sound, and resilient; and which builds the capabilities and skills necessary for future generations.
Committee on World Food Security, 2012 [46]	Individual	When all people, at all times, have physical, social, and economic access to sufficient, safe, and nutritious food that meets their dietary needs and food preferences for an active and healthy life ... The four pillars of food security are availability, access, utilization, and stability. The nutritional dimension is integral to the concept of food security.
Winne et al., 2000, p. 4 [59]	Regional	All persons in a community have access to culturally acceptable, nutritionally adequate food through local non-emergency sources at all times.
World Food Program, 2009, p. 170 [67]	Individual	A condition that exists when all people, at all times, are free from hunger.
World Food Summit, 1996 [2]	Individual	When all people, at all times, have physical and economic access to sufficient, safe, and nutritious food that meets their dietary needs and food preferences for an active and healthy life.

Note: While not all-inclusive, the definitions presented illustrate the need for a comprehensive definition. To our knowledge, all definitions are reported verbatim and are cited with their original source.

Table A2. Definitions of Food Insecurity Used in the Literature Since 2000.

Citation	Level of Analysis	Definition
Alaimo, 2005 [18]	Global, National, Regional	Limited or uncertain availability of nutritionally adequate or safe foods.
Babu and Gajanan, 2021 [31]	Individual	A lack of access to the kinds and amounts of food necessary for each member of a household to lead an active and a healthy lifestyle.
Bergmans, 2019 [92]	Individual	The physical pain of hunger as well as the more common experience of worrying about having enough healthy food to eat.
Bovell et al., 2015 [93]	Household	Limited or uncertain access to enough food for all household members to live active and healthy lives.
Coleman-Jensen et al., 2014 [94]	Household	A household-level economic and social condition of limited or uncertain access to adequate food.
Coleman-Jensen et al., 2021 [61]	Individual	Access to adequate food is limited by a lack of money and other resources.

Table A2. *Cont.*

Citation	Level of Analysis	Definition
Davitt et al., 2021 [95]	Household	Diminished variety, quality, and desirability of diet as well as decreased access to food.
Donley and Gualtieri, 2015 [96]	Individual	Lacking enough money to buy the amount and variety of food one needs or wants.
Dowler et al., 2001 [97]	Individual	The inability to acquire or consume an adequate quality or sufficient quantity of food in socially acceptable ways, or the uncertainty that one will be able to do so.
Economic Research Service, 2022 [28]	Household	Households were, at times, unable to acquire adequate food for one or more household members because the household had insufficient money and other resources for food.
FAO et al. 2022 [3]	Individual	Lack regular access to enough safe and nutritious food for normal growth and development and an active and healthy life. This may be due to unavailability of food and/or lack of resources to obtain food.
Food Forward, 2019 [98]	Household, Individual	Lack of access to enough, good, healthy, and culturally appropriate food.
Frongillo and Horan, 2004, p. 28 [47]	Global, National, Regional, Household, Individual	Problems with the availability, accessibility, and utilization of food.
Gundersen and Ziliak, 2015, p. 1830 [13]	Household	A condition which households lack access to adequate food because of limited money or other resources.
Harke et al., 2021, p. 1 [99]	Household, Individual	The lack of access to sufficient food because of limited financial resources.
Life Sciences Research Office and Andersen, 1990 [24]	National, Regional, Household	The limited or uncertain availability of nutritionally adequate and safe foods, or limited or uncertain ability to acquire acceptable foods in socially acceptable ways.
Litton and Beavers, 2021 [100]	Individual	Reduced quality, variety, or desirability of diet.
Mardones et al., 2020 [101]	Individual	Unreliable physical, social, and economic access to sources of adequate and nutritious food that meets people's dietary needs and food preferences.
Maxwell et al., 1990 [102]	Individual	The lack of access to enough food.
Miriam Webster Dictionary, 2022 [30]	Household, Individual	The fact or an instance of being unable to consistently access or afford adequate food.
National Research Council et al., 2006 [8]	Household, Individual	Individuals and/or families in a household adjusting their dietary intakes or preferences because of a lack of physical or economic resources.
National Research Council, 2006, p. 4 [8]	Household, Individual	Uncertain, insufficient, or unacceptable availability, access, or utilization of food.
National Research Council, 2006, p. 44 [8]	Individual	The social and economic problem of lack of food due to resource or other constraints, not voluntary fasting, or dieting, or because of illness, or for other reasons.
Niles et al., 2020 [103]	Household, Individual	The lack of consistent physical, social, and economic access to adequate and nutritious food that meets dietary needs and food preferences.
Nord and Prell, 2007 [104]	Household	Household level economic and social condition of limited access to food.

Table A2. *Cont.*

Citation	Level of Analysis	Definition
Nord et al., 2005 [105]	Household, Individual	The disruption of food intake or eating patterns because of lack of money and other resources.
Power et al., 2016, p. 4 [106]	Household, Individual	Inadequate or insecure access to adequate food due to financial constraints.
Roshanafshar and Hawkins, 2015, p. 4 [107]	Household	When one or more (household) members do(es) not have access to the variety or quantity of food that they need due to lack of money.
Shapouri, 2010, p. v [108]	Individual	Consuming less than the nutritional target of 2100 calories per day per person.
United Health Foundation, 2022 [17]	Household	Households unable to provide adequate food for one or more household members due to lack of resources.
US Department of Agriculture, 2022 [27]	Household	Lack of available financial resources for food at the household level.
US Department of Agriculture, 2022 [27]	Household	A household's inability to provide enough food for each person to live an active healthy lifestyle.
Voices for Alabama's Children, 2015 [109]	Household	Lack of access, at times, to enough food for an active, healthy life for all household members and limited or uncertain availability of nutritionally adequate foods.

Note: While not all-inclusive, the definitions presented illustrate the need for a comprehensive definition. To our knowledge, all definitions are reported verbatim and are cited with their original source.

Table A3. Different Ideas of 'Adequate Food'.

Citation	Definition
Dietary Needs	
Shapouri, 2010, p. v [108]	Consuming less than the nutritional target of *2100 calories per day* per person.
Nutritious (and Safe)	
Alaimo, 2005 [18]	Food insecurity: limited or uncertain availability of *nutritionally adequate* or *safe foods*.
Life Sciences Research Office and Andersen, 1990 [24]	Food insecurity: whenever the availability of *nutritionally adequate and safe foods* or the ability to acquire acceptable foods in socially acceptable ways is limited or uncertain.
Blumberg et al., 1999, p. 1231 [79]	Assured access to *nutritionally adequate* and *safe* foods without resorting to emergency food supplies, scavenging, stealing, and other coping strategies.
Wood et al., 2000 [88]	The state in which all persons obtain a *nutritionally adequate*, culturally acceptable diet at all times through local, non-emergency sources.
Life Sciences Research Office and Andersen, 1990 [24]	Access by all members at all times to enough food for an active, healthy life. Food security includes at a minimum: the ready availability of *nutritionally adequate* and *safe* foods; assured ability to acquire acceptable foods in socially acceptable ways (i.e., without resorting to emergency food supplies, scavenging, stealing, or other coping strategies).
Quality	
Prifti, 2021, p. 238 [72]	A function of availability of adequate food in terms of quantity and *quality* and the people's ability to afford it at all times.

Table A3. *Cont.*

Citation	Definition
Multifaceted	
Winne et al., 2000, p. 4 [59]	All persons in a community have access to *culturally acceptable, nutritionally adequate* food through local non-emergency sources at all times.
Committee on World Food Security, 2012 [46]	When all people at all times have physical, social, and economic access to food, which is *safe* and consumed in sufficient quantity and *quality* to meet their *dietary needs* and *food preferences*, and is supported by an environment of adequate sanitation, health services and care, allowing for a healthy and active life.
Davitt et al., 2021 [94]	Food insecurity: diminished *variety, quality, and desirability* of diet as well as decreased access to food.
Donley and Gualtieri, 2015 [95]	Food insecurity: lacking enough money to buy the amount and *variety* of food one *needs* or *wants.*
FAO, 2009 [2]	Food security exists when all people, at all times, have physical, social, and economic access to *sufficient, safe,* and *nutritious* food that meets their *dietary needs* and *food preferences* for an active and healthy life.
Food Forward, 2019 [97]	Food insecurity: lack of access to enough, *good*, healthy, and *culturally appropriate food.*
Niles et al., 2020 [103]	Food insecurity: the lack of consistent physical, social, and economic access to adequate and *nutritious* food that meets *dietary needs* and *food preferences.*
World Food Summit, 1996 [2]	When all people, at all times, have physical, social, and economic access to *sufficient, safe,* and *nutritious* food that meets their *dietary needs* and *food preferences* for an active and healthy life.

Table A4. Different Ideas of 'Preferred Foods'.

Citation	Definition
Tastes and Preferences	
Mardones et al., 2020 [101]	Food insecurity: unreliable physical, social, and economic access to sources of adequate and nutritious food that meets people's dietary needs and *food preferences.*
Niles et al., 2020 [103]	The lack of consistent physical, social, and economic access to adequate and nutritious food that meets dietary needs and *food preferences.*
Desirability	
Davitt et al., 2021 [95]	Food insecurity: diminished variety, quality, and *desirability of diet* as well as decreased access to food
Donley and Gualtieri, 2015 [96]	Food insecurity: lacking enough money to buy the amount and variety of food one needs or *wants.*
Litton and Beavers, 2021 [100]	Food insecurity: reduced quality, variety, or *desirability of diet.*
Culturally Acceptable	
Winne et al., 2000, p. 4 [59]	All persons in a community have access to *culturally acceptable,* nutritionally adequate food through local non-emergency sources at all times.
Wood et al., 2000 [89]	The state in which all persons obtain a nutritionally adequate, *culturally acceptable diet* at all times through local, non-emergency sources.
Food Forward, 2019 [98]	Food insecurity: lack of access to enough, *good*, healthy, and *culturally appropriate food.*

Table A5. Different Ideas of 'Culturally or Socially Acceptable Means'.

Citation	Definition
Non-Emergency Means	
Winne et al., 2000, p. 4 [59]	All persons in a community have access to culturally acceptable, nutritionally adequate food through local *non-emergency sources* at all times.
Blumberg et al., 1999, p. 1231 [79]	Assured access to nutritionally adequate and safe foods *without resorting to emergency food supplies, scavenging, stealing, and other coping strategies.*
Wood et al., 2000 [89]	The state in which all persons obtain a nutritionally adequate, culturally acceptable diet at all times through *local, non-emergency sources.*
Life Sciences Research Office and Andersen, 1990 [24]	Access by all members at all times to enough food for an active, healthy life. Food security includes at a minimum:the ready availability of nutritionally adequate and safe foods; assured ability to acquire acceptable foods in socially acceptable ways (i.e., *without resorting to emergency food supplies, scavenging, stealing, or other coping strategies*).
Monetary Means	
United Health Foundation, 2022 [17]	Food insecurity: households unable to provide adequate food for one or more household members due to *lack of resources,*
Coleman-Jensen et al., 2021 [61]	Access to adequate food is limited by a *lack of money and other resources.*
Donley and Gualtieri, 2015 [96]	Food insecurity: lacking enough *money to buy* the amount and variety of food one needs or wants.
Economic Research Service, 2022 [28]	Food insecurity: household members were, at all times, unable to acquire adequate food for one or more household members because the household had *insufficient money and other resources for food.*
Harke et al., 2021, p. 1 [99]	The lack of access to sufficient food because of *limited financial resources.*
Gundersen and Ziliak, 2015, p. 1830 [13]	Food insecurity: a condition which households lack access to adequate food because of *limited money or other resources.*
Office of Disease Prevention and Health Promotion, 2020 [110]	Food insecurity: the disruption of food intake or eating patterns because of *lack of money and other resources.*
USDA, 2021 [111]	Food insecurity: lack of available *financial resources for food* at the household level.

References

1. Maxwell, D.G. Measuring food insecurity: The frequency and severity of "coping strategies". *Food Policy* **1996**, *21*, 291–303. [CrossRef]
2. FAO; IFAD; UNICEF; WFP; WHO. *The State of Food Insecurity in the World: Economic Crises-Impacts and Lessons Learned*; FAO: Rome, Italy, 2009.
3. FAO; IFAD; UNICEF; WFP; WHO. *The State of Food Security and Nutrition in the World 2022: Repurposing Food and Agricultural Policies to Make Healthy Diets More Affordable*; FAO: Rome, Italy, 2022.
4. FAO; IFAD; UNICEF; WFP; WHO. *The State of Food Security and Nutrition in the World: Transforming Food Systems for Food Security, Improved Nutrition, and Affordable Healthy Diets for All*; FAO: Rome, Italy, 2021.
5. Jones, A.D.; Ngure, F.M.; Pelto, G.; Young, S.L. What are we assessing when we measure food security? a compendium and review of current metrics. *Adv. Nutr.* **2013**, *4*, 481–505. [CrossRef] [PubMed]
6. Barrett, C.B. Measuring Food Insecurity. *Science* **2010**, *327*, 825–828. [CrossRef] [PubMed]
7. Kendall, A.; Olson, C.M.; Frongillo, E.A., Jr. Relationship of hunger and food insecurity to food availability and consumption. *J. Am. Diet. Assoc.* **1996**, *96*, 1019–1024. [CrossRef]
8. National Research Council; Wunderlich, G.S.; Norwood, J.L. (Eds.) *Food Insecurity and Hunger in the United States: An Assessment of the Measure*; National Academies Press: Washington, DC, USA, 2006.
9. Maxwell, S.; Smith, M. Household food security: A conceptual review. *Househ. Food Secur. Concepts Indic. Meas.* **1992**, *1*, 1–72.

10. Shaw, D.J. *World Food Security, a History Since 1945*; Palgrave Macmillan: London, UK, 2007.
11. Erokhin, V.; Gao, T. Impacts of COVID-19 on trade and economic aspects of food security: Evidence from 45 developing countries. *Int. J. Environ. Res. Pub. Health* **2020**, *17*, 5775. [CrossRef] [PubMed]
12. Vozoris, N.T.; Tarasuk, V.S. Household Food Insufficiency Is Associated with Poorer Health. *J. Nutr.* **2003**, *133*, 120–126. [CrossRef]
13. Gundersen, C.; Ziliak, J.P. Food insecurity and health outcomes. *Health Aff.* **2015**, *34*, 1830–1839. [CrossRef]
14. O'Dare, W.K. Community Food Environments and Healthy Food Access Among Older Adults: A Review of the Evidence for the Senior Farmers' Market Nutrition Program (SFMNP). *Soc. Work Health Care* **2017**, *56*, 227–243. [CrossRef]
15. Bahn, R.A.; Hwalla, N.; Labban, S.E. Leveraging nutrition for food security: The integration of nutrition in the four pillars of food security. In *Food Security and Nutrition*; Academic Press: Chicago, IL, USA, 2021; pp. 1–32.
16. Akbari, M.; Foroudi, P.; Shahmoradi, M.; Padash, H.; Parizi, Z.S.; Khosravani, A.; Cuomo, M.T. The evolution of food security: Where are we now, where should we go next? *Sustainability* **2022**, *14*, 3634. [CrossRef]
17. United Health Foundation. Public Health Impact: Food Insecurity. 2022. Available online: https://www.americashealthrankings.org/explore/health-of-women-and-children/measure/food_insecurity_household (accessed on 20 March 2023).
18. Alaimo, K. Food insecurity in the United States: An overview. *Top. Clin. Nutr.* **2005**, *20*, 281–298. [CrossRef]
19. Widome, R.; Neumark-Sztainer, D.; Hannan, P.J.; Haines, J.; Story, M. Eating when there is not enough to eat: Eating behaviors and perceptions of food among food-insecure youths. *Am. J. Public Health* **2009**, *99*, 822–828. [CrossRef] [PubMed]
20. Gundersen, C.; Ziliak, J.P. Childhood food insecurity in the US: Trends, causes, and policy options. *Future Child* **2014**, 1–19. Available online: https://www.jstor.org/stable/26304695 (accessed on 1 March 2023). [CrossRef]
21. Weis, T. Food security and food sovereignty. In *International Encyclopedia of Human Geography*; Elsevier: Amsterdam, The Netherlands, 2020; Volume 2, pp. 191–196.
22. United Nations. Peace, Dignity, and Equality on a Healthy Planet. 2022. Available online: https://www.un.org/en/global-issues/food (accessed on 20 March 2023).
23. Maletta, H.E. From hunger to food security: A conceptual history. *Soc. Sci. Res. Netw.* **2014**, 2484166. [CrossRef]
24. Anderson, S.A. Core indicators of nutritional state for difficult-to-sample populations. *J. Nutr.* **1990**, *120*, 1555–1600. [CrossRef] [PubMed]
25. Olson, C.M.; Anderson, K.; Kiss, E.; Lawrence, F.C.; Seiling, S.B. Factors protecting against and contributing to food insecurity among rural families. *Fam. Econ. Nutr. Rev.* **2004**, *16*, 12–21.
26. Hayes, T.O. Food Insecurity and Food Insufficiency: Assessing Causes and Historical Trends. 2021. Available online: https://www.americanactionforum.org/research/food-insecurity-and-food-insufficiency-assessing-causes-and-historical-trends/ (accessed on 20 March 2023).
27. US Department of Agriculture. Definitions of Food Security. 2022. Available online: https://www.ers.usda.gov/topics/food-nutrition-assistance/food-security-in-the-u-s/definitions-of-food-security/ (accessed on 20 March 2023).
28. Economic Research Service. Measurement. 2022. Available online: https://www.ers.usda.gov/topics/food-nutrition-assistance/food-security-in-the-u-s/measurement/#:~:text=Food%20Insecure.,-Households%20that%20report&text=The%20three%20least%20severe%20conditions,have%20money%20to%20get%20more (accessed on 20 March 2023).
29. Peng, W.; Berry, E.M. The concept of food security. In *The Encyclopedia of Food Security and Sustainability*; Elsevier: Amsterdam, The Netherland, 2019; Volume 2, pp. 1–7.
30. Merriam-Webster. Food Insecurity. 2022. Available online: https://www.merriam-webster.com/dictionary/food%20insecurity (accessed on 20 March 2023).
31. Babu, S.C.; Gajanan, S.N.; Sanyal, P.; United Nations. Introduction to food security: Concepts and measurement. In *Food Security, Poverty and Nutrition Policy Analysis*; Academic Press: San Diego, CA, USA, 2009; pp. 5–15.
32. FAO. *World Food and Agriculture Situation*; FAO: Rome, Italy, 1974.
33. Maxwell, S.; Frankenberger, T.R. *Household Food Security: Concepts, Indicators, Measurements: A Technical Review*; UNICEF: New York City, NY, USA, 1995.
34. Bickel, G.; Nord, M.; Price, C.; Hamilton, W.; Cook, J. *Guide to Measuring Household food Security, Revised 2000*; U.S. Department of Agriculture, Food and Nutrition Service: Alexandria, VA, USA, 2000.
35. Smith, M.; Pointing, J.; Maxwell, S. *Household Food Security: Concepts and Definitions: An Annotated Bibliography*; Institute of Development Studies: Sussex, UK, 1993; Volume 8.
36. *The Rome Declaration on World Food Security & World Food Summit Plan of Action*; World Food Summit: Rome, Italy, 2009.
37. Clapp, J.; Moseley, W.G.; Burlingame, B.; Termine, P. The case for a six-dimensional food security framework. *Food Policy* **2021**, *106*, 102164. [CrossRef]
38. Beacom, E.; Furey, S.; Hollywood, L.; Humphreys, P. Perspectives on Food Poverty Definition and Measurement in Northern Ireland. 2018. Available online: https://www.researchgate.net/profile/Emma-Beacom-2/publication/344025192_Perspectives_on_food_poverty_definition_and_measurement_in_Northern_Ireland/links/605db20992851cd8ce69c965/Perspectives-on-food-poverty-definition-and-measurement-in-Northern-Ireland (accessed on 20 March 2023).
39. Coates, J. Build it back better: Deconstructing food security for improved measurement and action. *Glob. Food Secur.* **2013**, *2*, 188–194. [CrossRef]
40. Koczorowski, C.Z. Unpacking food security: Is there shelf space for food security studies? *Inq. Insight* **2010**, *3*, 39–62.
41. Dery, D.; Wildavsky, A.B. *Problem Definition in Policy Analysis*; University Press: Kansas, MI, USA, 1984.

42. Hay, R.W.; Rukuni, M. SADCC food security strategies: Evolution and role. *World Dev.* **1988**, *16*, 1013–1024. [CrossRef]
43. Fitzpatrick, L.E. Food Security and Theories of Development. Ph.D. Thesis, The University of Notre Dame, Notre Dame, IN, USA, 1997.
44. Lang, T.; Barling, D. Food security and food sustainability: Reformulating the debate. *Geogr. J.* **2012**, *178*, 313–326. [CrossRef]
45. Scott, K.W. Obesisty in the South: Do Subjective Measures of Food Access Matter? Ph.D. Thesis, University of Georgia, Athens, GA, USA, 2015.
46. Committee on World Food Security. Global Strategic Framework for Food Security and Nutrition: First Version. 2012. Available online: http://www.fao.org/docrep/meeting/026/ME498E.pdf (accessed on 20 March 2023).
47. Frongillo, E.A.; Horan, C.M. Hunger and aging. *Generations* **2004**, *28*, 28–33.
48. HLPE. Food Security and Nutrition: Building a Global Narrative towards 2030. A Report by the High Level Panel of Experts on Food Security and Nutrition of the Committee on World Food Security. 2020. Available online: https://www.fao.org/3/ca9731en/ca9731en.pdf (accessed on 20 March 2023).
49. Leroy, J.L.; Ruel, M.; Frongillo, E.A.; Harris, J.; Ballard, T.J. Measuring the food access dimension of food security: A critical review and mapping of indicators. *Food Nutr. Bull.* **2015**, *36*, 167–195. [CrossRef]
50. Sen, A. *Poverty and Famines: An Essay on Entitlement and Deprivation*; Oxford University Press: Oxford, UK, 1983.
51. Alamgir, M.; Arora, P. Providing Food Security for All. 1991. Available online: https://practicalactionpublishing.com/book/1736/providing-food-security-for-all (accessed on 20 March 2023).
52. IFAD. *Food Security in Africa [Paper Presentation]*; Symposium of Food Security: Dakar, Senegal, 1991; pp. 3–4.
53. Siamwalla, A.; Valdes, A. Food insecurity in developing countries. *Food Policy* **1980**, *5*, 258–272. [CrossRef]
54. World Bank Staff. *Food Security in Food-Deficit Countries*; [Working Paper]; World Bank: Washington, DC, USA, 1980.
55. Barraclough, S.L.; Utting, P. Food Security Trends and Prospects in Latin America. 1987. Available online: https://kellogg.nd.edu/sites/default/files/old_files/documents/099_0.pdf (accessed on 20 March 2023).
56. Benson, C.; Clay, E.J.; Green, R.H. *Food Security in Sub-Saharan Africa*; IDS, University of Sussex: Brighton, UK, 1986.
57. Eicher, C.K.; Staatz, J.M. Food security policy in sub-Saharan Africa in Agriculture in a Turbulent World Economy. In Proceedings of the International Conference on Agricultural Economics, Malaga, Spain, 26 August 1985.
58. Jonsson, U.; Toole, D. *Household Food Security and Nutrition: A Conceptual Analysis*; United Nations Children's Fund: New York, NY, USA, 1991.
59. Winne, M.; Hugh, J.; Fisher, A. (Eds.) *Community Food Security: A Guide to Concept, Design, and Implementation*; The Community Food Security Coalition: Milwaukee, WI, USA, 2000.
60. Bartfeld, J.; Dunifon, R. State-level predictors of food insecurity among households with children. *J. Assoc. Public Policy Anal. Manag.* **2006**, *25*, 921–942. [CrossRef]
61. Coleman-Jensen, A.; Rabbitt, M.P.; Gregory, C.A.; Singh, A. Household Food Security in the United States in 2020. 2021. Available online: https://www.ers.usda.gov/publications/pub-details/?pubid=102075 (accessed on 20 March 2023).
62. Haddad, L.; Chung, K.; Devi, P.Y. Alternative approaches to locating the food and nutrition insecure: Work in progress in South India. *Econ. Political Wkly.* **1995**, *30*, 392–402.
63. Kabeer, N. Women, Household Food Security, and Coping Strategies. 1990. Available online: https://agris.fao.org/agris-search/search.do?recordID=XF9109055 (accessed on 20 March 2023).
64. Reutlinger, S. Food security and poverty in LDCs. *Financ. Dev.* **1985**, *22*, 7–11.
65. United Nations. *Nutrition-Relevant Actions in the Eighties: Some Experience and Lessons from Developing Countries*; [Background paper]; Policies to Alleviate Undercompensation and Malnutrition in Deprived Areas: London, UK, 1990.
66. Reutlinger, S. *Poverty and Hunger: Issues and Options for Food Security in Developing Countries. A World Bank Policy Study*; The World Bank: Washington, DC, USA, 1986.
67. World Food Program. *Hunger and Markets*; Earthscan, 2009; Volume 3. Available online: https://www.wfp.org/publications/world-hunger-series-hunger-and-markets (accessed on 20 March 2023).
68. Ballard, T.J.; Kepple, A.W.; Cafiero, C. *The Food Insecurity Experience Scale: Development of a Global Standard for Monitoring Hunger Worldwide*; FAO: Rome, Italy, 2013.
69. Webb, P.; Coates, J.; Frongillo, E.A.; Rogers, B.L.; Swindale, A.; Bilinsky, P. Measuring Household Food Insecurity: Why it's so Important and Yet so Difficult to Do. *J. Nutr.* **2006**, *136*, 1404S–1408S. [CrossRef]
70. Vhurumuku, E. Food Security Indicators [PowerPoint]. Presented at the Integrating Nutrition and Food Security Programming for Emergency Response Workshop. 2014. Available online: https://www.fao.org/fileadmin/user_upload/food-security-capacitybuilding/docs/Nutrition/NairobiWorkshop/5.WFP_IndicatorsFSandNutIntegration.pdf (accessed on 26 July 2022).
71. McCordic, C.; Frayne, B.; Sunu, N.; Williamson, C. The Household Food Security Implications of Disrupted Access to Basic Services in Five Cities in the Global South. *Land* **2022**, *11*, 654. [CrossRef]
72. Prifti, E. *Income Versus Prices: How Does the Business Cycle Affect Food (in)-Security?* IMF Working Papers, 238; International Monetary Fund: Washington, DC, USA, 2021.
73. Gustafson, D.; Gutman, A.; Leet, W.; Drewnowski, A.; Fanzo, J.; Ingram, J. Seven Food System Metrics of Sustainable Nutrition Security. *Sustainability* **2016**, *8*, 196. [CrossRef]
74. Lentz, E.; Bageant, E.; Narayanan, S. Empowerment and nutrition in Niger: Insights from the women's empowerment in nutrition grid. *Food Secur.* **2021**, *13*, 1227–1244. [CrossRef]

75. Narayanan, S.; Lentz, E.; Fontana, M.; Kulkarni, B. Rural women's empowerment in nutrition: A framework linking food, health and institutions. *J. Dev. Stud.* **2022**, *58*, 1–18. [CrossRef]
76. Cafiero, C.; Viviani, S.; Nord, M. Food Security Measurement in a Global Context: The Food Insecurity Experience Scale. *Measurement* **2018**, *116*, 146–152. [CrossRef]
77. Watson, H. What is Global Food Insecurity? *The Borgen Project*. 2013. Available online: https://borgenproject.org/what-is-global-food-insecurity/ (accessed on 21 June 2022).
78. Food and Agriculture Organization of the United Nations. Hunger and Food Insecurity. 2022. Available online: https://www.fao.org/hunger/en/ (accessed on 20 March 2023).
79. Blumberg, S.J.; Bialostosky, K.; Hamilton, W.L.; Briefel, R.R. The effectiveness of a short form of the household food security scale. *Am. J. Public Health* **1999**, *89*, 1231–1234. [CrossRef]
80. International Dietary Data Expansion Project. Data4Diets: Building Blocks for Diet-Related Food Security Analysis. 2018. Available online: https://inddex.nutrition.tufts.edu/data4diets (accessed on 29 June 2022).
81. Roser, Max and Hannah Ritchie. Hunger and Undernourishment. 2019. Available online: https://ourworldindata.org/hunger-and-undernourishment (accessed on 21 June 2022).
82. Liang, Y.; Zhong, T.; Crush, J. Boon or Bane? Urban Food Security and Online Food Purchasing during the COVID-19 Epidemic in Nanjing, China. *Land* **2022**, *11*, 945. [CrossRef]
83. Coates, J.; Frongillo, E.A.; Rogers, B.L.; Webb, P.; Wilde, P.E.; Houser, R. Commonalities in the experience of household food insecurity across cultures: What are measures missing? *J. Nutr.* **2006**, *136*, 1438S–1448S. [CrossRef]
84. Feyertag, J.; Munden, L.; Calow, R.; Gelb, S.; McCord, A.; Mason, N. COVID-19 and Food Security: Novel Pathways and Data Collection to Improve Research and Monitoring Efforts. 2021. Available online: www.odi.org/en/publications/covid-19-and-food-security-novel-pathways-and-datacollection-to-improve-research-and-monitoring-efforts (accessed on 20 March 2023).
85. Hamad, H.; Khashroum, A. Household Food Insecurity (HFIS): Definitions, Measurements, Socio-Demographic and Economic Aspects. *J. Nat. Sci. Res.* **2016**, *6*, 63–75.
86. Simon, G.A. *Food Security*; University of Roma Tre: Rome, Italy, 2012.
87. Watuleke, J. *The Role of Food Banks in Food Security in Uganda: The Case of the Hunger Project Food Bank, Mbale Epicentre*; Current Afican. Issues; Nordiska Afrikainstitutet: Uppsala, Sweden, 2015; Volume 60.
88. Béné, C. Resilience of local food systems and links to food security: A review of some important concepts in the context of COVID-19 and other shocks. *Food Secur.* **2020**, *12*, 805–822. [CrossRef]
89. Wood, B.; Ross, K.; Kouris-Blazos, A. *Healthy Eating for Nutritionally Vulnerable Groups: Some Health Promotion Initiatives*; Healthy Eating Healthy Living Program, Monash University: Melbourne, Australia, 2000.
90. Siche, R. What is the impact of COVID-19 disease on agriculture? *Sci. Agropecu.* **2020**, *11*, 3–6. [CrossRef]
91. Sustainable Development Commission. Food Security and Sustainability: The Perfect Fit. 2009. Available online: https://www.sd-commission.org.uk/data/files/publications/SDCFoodSecurityPositionPaper.pdf (accessed on 20 March 2023).
92. Bergmans, R.S. Food insecurity transitions and smoking behavior among older adults who smoke. *Prevent. Med.* **2019**, *126*, 105784. [CrossRef]
93. Bovell, A.R.; Ettinger de Cuba, S.; Casey, P.H.; Coleman, S.; Cook, J.T.; Cutts, D.; Heeren, T.C.; Meyers, A.; Sandel, M.; Black, M.M.; et al. Trends in household and child food insecurity among families with young children from 2007 to 2013. *J. Appl. Res. Child. Inf. Policy Child. Risk* **2015**, *6*, 3.
94. Coleman-Jensen, A.; Gregory, C.; Singh, A. Household Food Security in the United States in 2013; USDA-ERS Economic Research Report 173. 2014. Available online: www.ers.usda.gov/publications/err-economic-research-report/err173.aspx (accessed on 20 March 2023).
95. Davitt, E.D.; Heer, M.M.; Winham, D.M.; Knoblauch, S.T.; Shelley, M.C. Effects of COVID-19 on university student food security. *Nutrients* **2021**, *13*, 1932. [CrossRef] [PubMed]
96. Donley, A.M.; Gualtieri, M.C. Food Deserts in US Cities. *Int. Encycl. Soc. Behav. Sci.* **2015**, *2*, 285–289.
97. Dowler, E.; Turner, S.A.; Dobson, B. *Poverty Bites: FOOD, Health, and Poor Families*; Cpag: London, UK, 2001; Volume 105.
98. Food Forward. What is Food Insecurity? *Food Security?* 2019. Available online: https://foodforward.org/food-security/what-is-food-insecurity/ (accessed on 20 March 2023).
99. Harke, M.; Dewey, A.; Engelhard, E.; Strayer, M.; Dawes, S.; Summerfelt, T.; Gundersen, C. The Impact of the Coronavirus on Food Insecurity in 2020 & 2021. 2021. Available online: https://www.feedingamerica.org/sites/default/files/2021-03/National%20Projections%20Brief_3.9.2021_0.pdf (accessed on 20 March 2023).
100. Litton, M.M.; Beavers, A.W. The relationship between food security status and fruit and vegetable intake during the COVID-19 pandemic. *Nutrients* **2021**, *13*, 712. [CrossRef] [PubMed]
101. Mardones, F.O.; Rich, K.M.; Boden, L.A.; Moreno-Switt, A.I.; Caipo, M.L.; Zimin-Veselkoff, N.; Alateeqi, A.M.; Baltenweck, I. The COVID-19 pandemic and global food security. *Front. Vet. Sci.* **2020**, *7*, 578508. [CrossRef] [PubMed]
102. Maxwell, S.; Swift, J.; Buchanan-Smith, M. Is Food Security Targeting Possible in Sub-Saharan Africa? Evidence from North Sudan. *IDS Bull.* **1990**, *21*, 52–61. [CrossRef]
103. Niles, M.T.; Bertmann, F.; Belarmino, E.H.; Wentworth, T.; Biehl, E.; Neff, R. The early food insecurity impacts of COVID-19. *Nutrients* **2019**, *12*, 2096. [CrossRef]

104. Nord, M.; Prell, M. Struggling to Feed the Family: What Does It Mean to Be Food Insecure? 2007. Available online: https://www ers.usda.gov/amber-waves/2007/june/struggling-to-feed-the-family-what-does-it-mean-to-be-food-insecure/ (accessed on 20 March 2023).
105. Nord, M.; Andrews, M.; Carlson, S. *Household Food Security in the United States, 2005*; (Report No. 29.); USDA Economic Research Service: Washington, DC, USA, 2005.
106. Power, E.; Abercrombie, D.; Fafard St-Germain, A.A.; Vanderkooy, P. *Prevalence, Severity, and Impact of Household Food Insecurity: A Serious Public Health Issue*; Dietitians of Canada: Toronto, ON, Canada, 2006; pp. 1–55.
107. Roshanafshar, S.; Hawkins, E. *Food Insecurity in Canada*; Statistics Canada: Ottawa, ON, Canada, 2015.
108. Shapouri, S. *Food Security Assessment, 2010–2020*; DIANE Publishing: Darby, PA, USA, 2010.
109. Voices for Alabama's Children. 2015. Available online: https://datacenter.kidscount.org/about/state-providers/details/1-voices-for-alabamas-children (accessed on 20 March 2023).
110. Office of Disease Prevention and Health Promotion. 2020. Available online: https://health.gov/healthypeople/priority-areas/social-determinants-health/literature-summaries/food-insecurity (accessed on 21 March 2023).
111. USDA. 2021. Available online: https://www.ers.usda.gov/topics/food-nutrition-assistance/food-security-in-the-u-s/measurement/ (accessed on 21 March 2023).

foods

MDPI

Review

Food Security Review Based on Bibliometrics from 1991 to 2021

Junfang Li [1,2] and Wei Song [1,3,*]

[1] Key Laboratory of Land Surface Pattern and Simulation, Institute of Geographic Sciences and Natural Resources Research, Chinese Academy of Sciences, Beijing 100101, China
[2] School of Earth Science and Resources, Chang'an University, Xi'an 710054, China
[3] Hebei Collaborative Innovation Center for Urban-Rural Integration Development, Shijiazhuang 050061, China
* Correspondence: songw@igsnrr.ac.cn

Abstract: Food security is related to human wellbeing and sustainable development and an important guarantee for world peace. In the context of global climate change, increased food demand, resource depletion, conflicts, and frequent public health emergencies, food security is widely seen as one of the top challenges. Food security research has obvious interdisciplinary characteristics, involving a wide range of fields. We analyzed the literature on food security in the Web of Science core collection from 1991 to 2021, using bibliometric methods with the aid of the Biblioshiny software package. By collecting, screening, analyzing, and visually expressing the literature data, the following conclusions were drawn: (1) In the past 30 years, the annual number of publications on food security increased. The period can be divided into three stages: 1991–2003 as the budding period, 2004–2012 as the development period, and 2013–2020 as the high-yield and active period. The top three journals discussing food security issues are *Food Security*, *Sustainability*, and *Food Policy*, and these journals focus on the publication of comprehensive views from interdisciplinary perspectives. (2) Studies on food security cover 138 countries or regions. The top three countries in terms of the number of published articles are the United States, the United Kingdom, and China. Among the top 20 countries in terms of the number of published articles, European countries are highlighted. (3) Climate change, food security, agriculture, policy, and management are the other high-frequency keywords in the field of food security; climate change occurred 321 times. The word sub-Saharan Africa also occurred more frequently, indicating that food security in sub-Saharan Africa has attracted wide attention. (4) The food security theme mapping clearly showed the research status and future development trends of various topics in the field. Currently, food production, climate change, and sustainable development are the most popular themes. Research on food sovereignty, ecological agriculture, child obesity, and other aspects is an emerging field. (5) We predict that in the future, the field of food security may focus on the expansion and improvement of the food security evaluation system, the balance between sustainable development and food security goals, the improvement of agricultural production and management efficiency, and the research on government policies and strategies. Our results provide a reference for grasping the current situation, key research direction, and development trend in the field of food security.

Keywords: food security; bibliometrics; sustainable development; thematic analysis

1. Introduction

Food security concerns the stability of the international community and human wellbeing and plays an important role in sustainable development. The 2030 Agenda for Sustainable Development lists hunger eradication, achieving food security, and improving nutritional status as high-priority areas in the 17 Sustainable Development Goals [1]. In recent years, with the frequent occurrence of wars, extreme weather events, and public security incidents, global food security issues have become the focus of attention [2,3]. The innovation and development of science and technology promoted grain production in countries around the world, but the huge population base still brings great challenges to the carrying capacity of food resources. The number of people globally affected by hunger increased from 768 million in 2020 to 828 million in 2021 [4], highlighting the worrying situation of global food security.

Currently, research on food security mainly includes the connotation of food security, the influencing factors, and how to guarantee food security. In 1974, the UN Food and Agriculture Organization first defined food security as "we should ensure that anyone anywhere can get enough food for survival and health in the future" [5] (pp. 303–320). The definition of food security, as agreed upon at the 1996 World Food Summit, contains four aspects: food, access, use, and stability. With the deepening of research, the concept of food security has been continuously refined and expanded, taking food hygiene, politics, human rights, and social culture into consideration. Therefore, food security not only refers to the problem of adequate food supply but also involves a balanced regional distribution and nutrition as well as stable eating habits. In 2012, the FAO updated the definition of food security, completing the development from only meeting survival needs to meeting positive and healthy living and food preferences [6]. The concept of food security from the initial emphasis on "food supply" to the later "focus on family and individual food acquisition power" indicates the extension from the macro overall to the micro individuals. After the 2012 revision, the concept of food security included macro- and micro-aspects, quantity and quality, and supply and demand. It is predicted that by 2050, food production needs to increase by 70% to meet the demands of the growing population. Currently, about 750 million people in the world face food security problems, and 1 billion people face long-term malnutrition [7].

Therefore, based on the macro-background of global climate change, wars and social conflicts, the uneven spatial distribution of water and soil resources, and the declining land production potential, several explorations from different perspectives and levels have been performed. Climate change is one of the important triggers of global food security, with complex impact mechanisms. The extent to which food insecurity can be attributed to climate change was quantitatively assessed by Dasgupta and Robinson [8]. Climate change will not only affect the global food supply directly by affecting food production but will also further push up food prices, which will exacerbate food security inequalities. Many efforts have been made at the national level to ensure the food security of certain countries. For example, in the face of food security threats caused by the decline in the rural population [9], rural aging [10,11], and abandoned farmland [12,13], China has actively adopted relevant policies and measures, such as strictly observing the red line of cultivated land (referring to the minimum land area for regular cultivation; China's current red line is 120 million hectares of cultivated land), the construction of high-standard farmland, and the comprehensive improvement of rural land [13,14]. Policies and the international environment have a significant impact on regional food security. Recently, the emergence of the COVID-19 pandemic, conflicts between Russia and Ukraine, and other crises have had a huge impact on international grain trade and sounded the alarm for some countries and regions that rely heavily on grain imports. For example, Singapore's food supply chain was seriously threatened [15] by these incidents. Although food security in Asia and Latin America has improved due to the development of science and technology, improved agricultural productivity, and government intervention, the problem of regional distribution inequality remains. At present, the food risk in some developing countries in Africa, South Asia, and other regions is still high [16].

On the research scale, food security has global, national, family, and individual aspects. At the national level, food security is deeply affected by the population base, the population growth rate, and the urbanization level. Countries at different stages of development have different priorities for ensuring food security. For example, developed countries focus on the issue of regional food security balance, whereas developing countries, such as Africa, are committed to eliminating the food threat posed by poverty [17]. National food security is an important guarantee for food security at the family level. Low family income, large gaps between rich and poor, and immature markets will all lead to family food insecurity.

Personal food security is highly important for personal nutrition and health, and excessive nutrient intake can cause many diseases, such as childhood obesity [18]. Analyzing the factors affecting food security is necessary to find adequate solutions. The influencing factors have been widely studied, and an evaluation index system has been established.

Research in the field of food security involves multiple disciplines and the comprehensive analysis of food security-related research from an interdisciplinary perspective to obtain an in-depth understanding of the current research situation. After Pritchard proposed the term bibliometrics in 1969, bibliometrics attracted global attention, which was most evident in the late 1970s. Bibliometric research mainly focuses on three fields: methodology, scientific information, and scientific policy, of which the first is the basic research field [19]. With the gradual maturity of bibliometric methods, this approach has been widely used in the quantitative analysis of literature information in various disciplines. In the process of practical application, it has been continuously expanded and extended, promoting the generation and development of scientific measurement methods, information measurement methods, and network information measurement methods. Because of the great significance of clarifying the development context of food security and putting forward the dual goals of food security and sustainable development, we conducted a quantitative analysis of the number of articles, authors, institutions, and keywords in the field of food security. The research questions were as follows:

(1) Which topics containing food are the most popular in academia?
(2) What is the current state of global cooperation on food security?
(3) How do the keywords in food security research cluster together?
(4) What areas of food security need further research?

2. Data Sources and Research Methods

2.1. Data Sources

Web of Science is the largest and most comprehensive collection of information resources in the world and contains more than 12,000 authoritative and high-impact academic journals in the fields of natural sciences, engineering, and biomedicine. We used the Web of Science core collection from the Web of Science database as the data source; the search method was the Topic Subject (TS) search, and the language was English. The search formula is TS = ("Food Security" or "Grain Security"). After deduplication and screening, 3734 documents in the field of food security, published between 1991 and 2021, were obtained, including 2832 research papers; 64 conference papers; 402 review articles; and 436 books, book chapters, letters, and other types of documents.

2.2. Research Methods

2.2.1. Methods of Bibliometric Analysis

Bibliometry, a traditional quantitative analysis method in the field of library and intelligence science, originated during World War II. After a long development period, it has been used in many disciplines and has become an important analysis method in scientific research. We combined bibliometric and other measurement research methods to process the retrieved literature information. The resulting data were further explored and analyzed using the software packages Bibliometric and VOS viewer and finally visualized for expression. This paper mainly observed and interpreted the overall literature data in the field of food security based on the publication volume evolution trend analysis, historical citation analysis, popular journal analysis, high-yield author analysis, main research country/institution analysis, keyword analysis, and theme analysis.

2.2.2. Research Ideas

The first step in a research review is conducting a research design, including raising research questions, clarifying research objectives, and identifying research methods (Figure 1). The second step is data collection and screening, using the literature research method for data collection, screening, and processing after collection. The third step is data analysis and visual expression. We first imported the data into Bibliometric for analysis, analyzed the input of food security research subjects (authors, journals, institutions, among others), and then discussed and analyzed the research hot-spots and preface frontiers combined with VOS viewer.

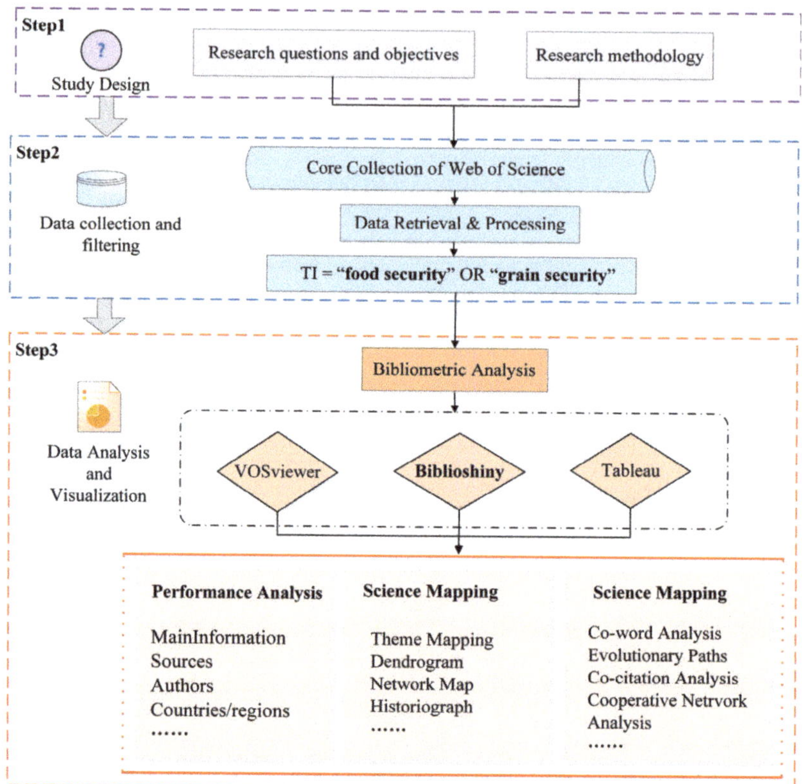

Figure 1. Measurement and analysis design process in the field of food security.

3. Analysis of Results

3.1. Evolution Trend of Publication Volume

The number of documents is an important indicator of the development stage of a certain field and a reference standard to predict the future trend. Figure 2 shows the development trend of research in the field of food security based on an analysis of the distribution of annual publications. From 1991 to 2021, the overall number of food security documents increased, with an average annual growth rate of 23.68%. Food security research can be roughly divided into three stages: In the first stage (1991–2003), the number of publications was small and the growth was slow. The field of food security had not yet received wide attention from the academic community, and the related studies were mainly focused on qualitative analysis. However, some pioneering articles laid the foundation for food security research. For example, an article published in 1992 obtained 190 citations, with an average annual reference volume of 6.33 [20]. In the second stage (2004–2012), more scholars began to focus on food security involving cultural, political, social, and other aspects, and this field received considerable interdisciplinary attention. Scientists began to discuss food security issues from different angles, such as transgenic technology, bringing opportunities and challenges to food security. In the third stage (2013–2021), the number of articles increased rapidly, making this stage one of high yield and activity. Food security attracted wide attention and gradually became an important issue.

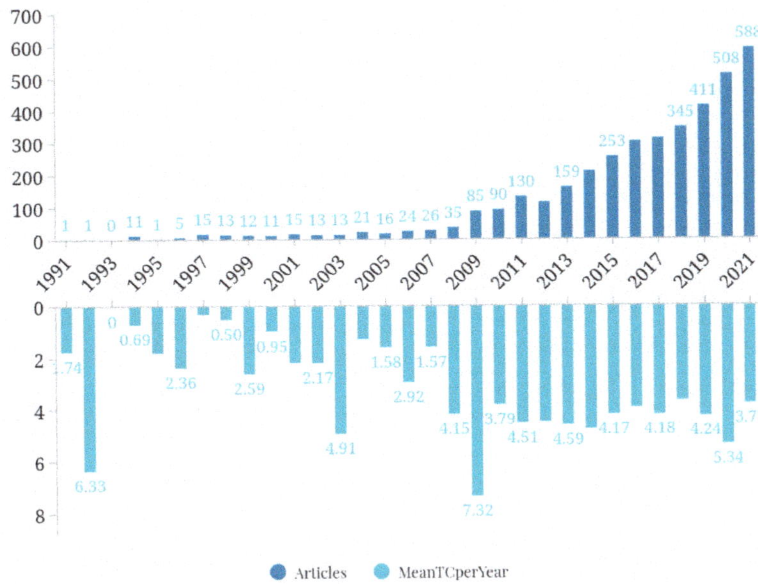

Figure 2. Number of published and cited food security research articles from 1991 to 2021.

3.2. Citation Analysis of Food Security Studies

By analyzing the local and global citation rates of the articles in the field of food security, we found that most of the top-cited articles were published around 2010, indicating that the research results in the field of food security during this period were more influential (Table 1). The article with the most local citations (LCs) was one by Professor Per Pinstrup-Andersen of Cornell University, published in *Food Security* in 2009, on the issues of "how to define food security" and "how to measure food security" [21]. The second-ranked article in terms of LCs synthesized the perspectives of the social and natural sciences; constructed a theoretical framework for the interaction of food security, ecosystem services, and social welfare; and revealed the key processes and main factors that controlled food through the elaboration of integrated systems of food production, supply, access, and use [22]. The establishment of this theoretical framework laid a foundation for future studies of the impact of global environmental changes on food security. The numbers of local citations of the above two articles both exceeded 100, indicating outstanding contributions in the field of food security. The third-ranked article in terms of LCs analyzed the definitions, conceptual frameworks, and household food security indicators of food security, defining food security as a global, national, family, and personal multifaceted concept. To better monitor and measure household food security, the advantages and disadvantages of different indicators at the specific level of analysis were evaluated [23]. Dana [24] called for global attention to be paid to phosphorus, an important raw material needed for food production, and suggested the inclusion of long-term phosphorus deficiency in the global food security priority agenda. The article obtained 2722 global citations, indicating that this research was widely recognized. Most of the top 10 articles in local citations were studies on the definition of food security [21], the conceptual framework [22], and the measurement index [23,25]. In addition, publications on food security impact factors, climate change and food security, and the impacts of food losses and waste on resources such as chemical fertilizers and fields [26,27] also attracted more attention.

Table 1. Top 10 articles in the field of food security.

Title	DOI	Year	LCs	GCs
Food security: definition and measurement	10.1007/s12571-008-0002-y	2009	103	419
Conceptualizing food systems for global environmental change research	10.1016/j.gloenvcha.2007.09.002	2008	101	556
Towards better measurement of household food security: Harmonizing indicators and the role of household surveys	10.1016/j.gfs.2012.11.006	2013	59	157
Food security: a post-modern perspective	10.1016/0306-9192(95)00074-7	1996	56	255
The story of phosphorus: Global food security and food for thought	10.1016/j.gloenvcha.2008.10.009	2009	45	2722
A food systems approach to researching food security and its interactions with global environmental change	10.1007/s12571-011-0149-9	2011	45	266
Rethinking the measurement of food security: from first principles to best practice	10.1007/s12571-013-0253-0	2013	45	131
Lost food, wasted resources: Global food supply chain losses and their impacts on freshwater, cropland, and fertiliser use	10.1016/j.scitotenv.2012.08.092	2012	44	570
Climate change and food Systems	10.1146/annurev-environ-020411-130608	2012	43	904
Does adaptation to climate change provide food security? A micro-perspective from Ethiopia	10.1093/ajae/aar006	2011	42	493

Note: Local citations (LCs), number of citations in papers in this article's database; global citations (GCs), number of citations in all papers.

In Biblioshiny, 20 nodes were selected, representing seminal work within the domain and some classical studies (Figure 3). The earliest node in the literature on food security was an article published in 1996 in *Food Policy*, entitled "Food security: a post-modern perspective" [28], with three different reference chains, ranking fourth with 56 LCs. This article identified three major shifts in food security thinking from the world and the country to families and individuals, from food first to livelihood, and from objective indicators to subjective perception. An interdisciplinary article by Dana [24], published in *Global Environmental Change*, revealed the possibility of a phosphate shortage in the future, emphasizing the importance of scientifically managing the phosphorus resources in the global food system.

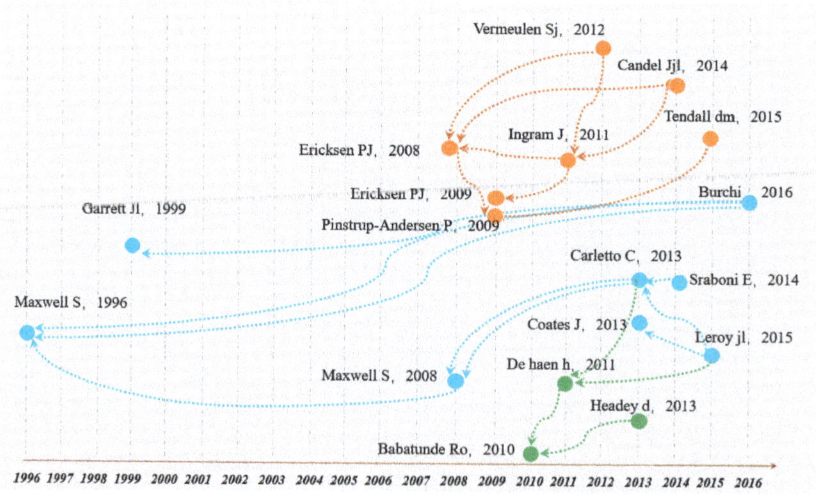

Figure 3. Historical direct citation network of papers published from 1996 to 2021 in the field of food security.

3.3. Popular Journal Analysis

Academic journals all have their own purpose, scope, focus, and research interests. Popular journals not only represent the recognition and choice of most researchers in this field but also are an important way to obtain hot research content. Researchers can also accurately locate their research results by interpreting articles published in popular journals. According to Bradford's Law, the concentration and dispersion of papers in the food security field from 1991 to 2021 were analyzed; the periodicals were arranged in descending order according to the number of papers published in the field of food security and divided into core, related, and unrelated areas. The numbers of journals in the core areas, related areas, and unrelated areas have the relationship 1: n: n^2. The core area was composed of seven journals; the journal with the largest number of articles was *Food Security*, and that with the highest H-index was *Food Policy* (Table 2).

Table 2. Top 7 journals published from 1991 to 2021 in the field of food security influence.

Journals	Articles	H-Index	TC	PY Start
Food Security	355	47	9701	2009
Sustainability	286	29	3685	2011
Food Policy	189	48	7453	1994
Global Food Security	155	39	3685	2008
World Development	105	36	4099	1997
International Journal of Environmental Research and Public Health	103	18	1179	2009
Frontiers in Sustainable Food Systems	85	10	410	2018

Note: H-index is a quantitative indicator of the number and level of academic output. The higher the value, the greater the impact of the journal in this field. TC is the total citations. PY Start is the year when the journal began to publish articles in the field of food security.

Food Security is an interdisciplinary international journal focused on exploring global food security issues. The journal pays special attention to the publication of a comprehensive view of food production, agricultural development, access to food and nutrition science, sociology, and economics. *Food Security* and *Sustainability* have seen a high growth rate of publications in recent years, and their cumulative publications surpassed those of *Food Policy* in 2014 and 2019, respectively, making them the most popular journals in the field (Figure 4).

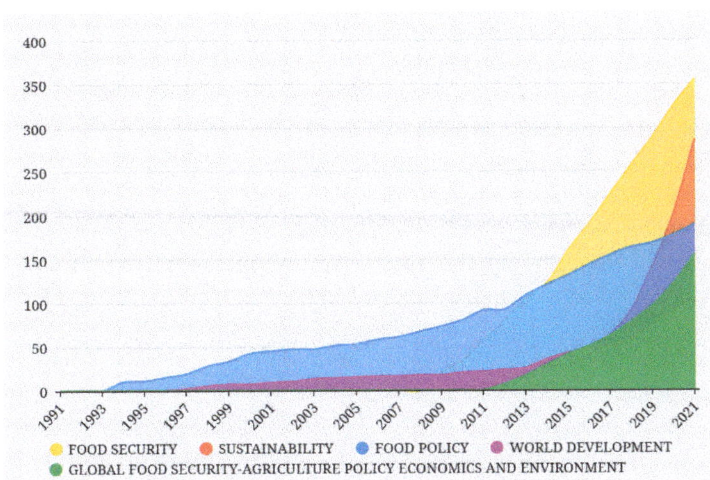

Figure 4. Popular journals in the field of food security from 1991 to 2021.

3.4. *Analysis of the Distribution Characteristics of the Major Research Countries and Institutions*

The dataset used for the analysis of food security studies contained publications from 134 countries/regions (Figure 5). European countries accounted for the majority of the top 20 countries, followed by countries in the Americas. The United States had most of the publications, with 1804 articles, about twice the number of the UK, which had the second highest number of publications. The number of US publications also increased more rapidly than that of the other countries, suggesting that the US was at the forefront of food security research. China ranked third in the number of articles, with 757 articles, with a high growth rate from 2020 to 2021. Because of the basic conditions of China, with a higher population and a relatively low land area, food security received high attention, with many publications in the fields of ensuring the ability to obtain food [29], forecasting the demand for food and nutrition security [30], and balancing the supply and demand structure of food [31].

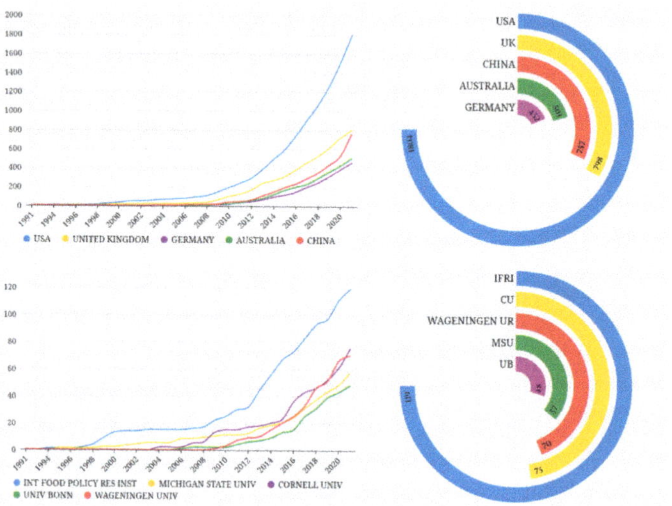

Figure 5. Top 10 countries and institutions publishing in the field of food security. Note: IFPRI: International Food Policy Research Institute; CU: Cornell University; WA-GENINGEN UR: Wageningen University & Research; MSU: Michigan State University; UB: Rheinische Friedrich-Wilhelms-Universität Bonn.

The agency with the greatest publication output is the International Food Policy Research Institute (IFPRI), with 119 articles. It is an international organization dedicated to providing research on food security, hunger eradication, and poverty and is a benchmark research institute in the field of food security. It is followed by Cornell University and Wageningen University, with 75 and 70 papers, respectively (Figure 5). The former and the latter are the world's top agricultural and environmental science research institutions and enjoy a high international reputation. The University of Michigan ranked fourth in publications and was responsible for some high-quality articles that were highly influential in the field. For example, an article on the assessment of food security indicators, published in *Advances in Nutrition* in 2013, received 339 citations [32]. The agencies that published the largest numbers of papers were mostly located in the United States, the Netherlands, and Germany.

3.5. *Keyword Analysis*

Keywords are a high-level summary of the topic of the article, and analyzing keywords in the field can quickly and accurately capture research hot-spots. Regarding the field of food security, the frequency of individual keywords represented the number of research findings. Climate change was the most frequently encountered keyword and was detected 321 times. Climate change directly affected the four pillars of food security (quantity, availability, usability, and stability) and had the most extensive impact on food security. The second most frequently occurring keywords were agriculture, policy, management, poverty, and insecurity, indicating that these aspects were highly related to food security. Sub-Saharan Africa (68 times), China (50 times), and India (27 times) appeared as keywords, showing that these countries were the focus of food security research (Figure 6). For a long time, the issue of food security in sub-Saharan Africa has received extensive attention from the international community. Substantial research was undertaken to identify existing problems in food security [33], eliminate regional poverty [14,34], and deal with the impact of climate change. Food security in China and India is also of concern due to the large population bases of these countries.

Figure 6. High-frequency keywords in the field of food security (top 50 keywords).

3.6. *Theme Analysis*

Theme mapping is a method that combines the advantages of traditional indexing and artificial intelligence, with keywords as the core of the data model to efficiently analyze its centrality and density, with the aim of exploring and reasoning about research topics. Theme maps can solve the problems caused by a large amount of disordered information. The horizontal and vertical axes indicate centrality and density, respectively, representing the workload and importance of the subject, respectively [35]. The theme map has four quadrants, and the current development stage of a theme can be analyzed by the location of the quadrant of the theme. Motor themes located in the first quadrant were both important and soundly trending, whereas niche themes in the second quadrant, although well developed, were of less importance to the field of food security. Emerging or declining themes in the third quadrant were marginal, with little workload and importance, whereas basic themes in the fourth quadrant were important for food security, with room for development (Figure 7).

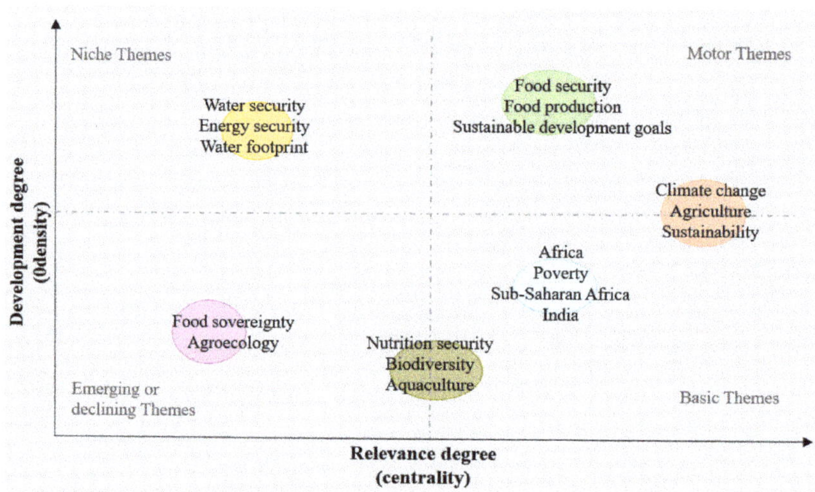

Figure 7. Distribution of themes in food security areas.

Food production, climate change, and sustainability are currently the most popular themes, with high research density and centrality in the field of food security. The topics related to water security, energy security, water footprint, and the environment, whilst receiving a lot of attention, are not closely integrated with food security. The number and centrality of studies on food sovereignty, ecological agriculture, and childhood obesity were not high, indicating that these are emerging fields. In recent years, due to the impact of international conflicts and the COVID-19 pandemic, the issue of food sovereignty has received extensive attention [36]. Research on nutrition security, biodiversity, and aquaculture had a medium degree of centrality, but the degree of intensity was not high, indicating that although these topics were closely related to food security, the number of published papers was small, and the future research potential was high. Sustainable agricultural development and poverty eradication in Africa, Southeast Asia, and other regions are the basic research directions in the field of food security.

4. Discussion

4.1. Analysis of the Research Trend of Food Security

Research on food security concerns various areas, and the content of such research has been constantly enriched, with significant progress in recent years. Studies in this field are of great significance for judging the development trend of food security scientifically and rationally. The development and evolution of the concept of food security have obvious stage and historical characteristics. The understanding of the connotation of food security has been constantly enriched and improved, and the results have gradually changed from theoretical to applied results. At first, food security mainly emphasized the sufficiency of food production and supply. After that, studies in food security began to pay equal attention to the total amount of food and quality safety and to the balance of food supply and demand [5,6]. In recent years, scholars have focused on various food security issues, from simple quantitative security to nutrition security, life security, and ecological security.

The whole research and development process in the field of food security can be understood through the analysis of article references, high-frequency keywords, countries and institutions, and thematic maps. At present, the mainstream view is that the countermeasures and suggestions for food security should be discussed mainly based on the four pillars, availability, access, utilization, and stability. Climate change, food nutrition and security, and sustainable agricultural development are hot issues in the field of food security research. From the perspective of disciplines, the interdisciplinary trend of food security research is obvious, and a growing number of scholars in environmental science, demography, and geography are paying attention to food security. Some research fields focus on ensuring food security and putting forward corresponding countermeasures and suggestions. Scientists construct the evaluation index system of food security from the perspective of food production, distribution, consumption, and food reserves [37]. Among the many factors affecting food security, population is the most direct one, and the number of people directly affects the food supply pressure. The issue of food security in India and sub-Saharan Africa has received extensive attention from the international community. Historical reasons, natural resource constraints, climate disasters, agricultural production technology, and management level are all factors impacting food security in Africa. Taking them into account and exploring their interrelationships and influences can be the key to ensuring food security in Africa. It is worth noting that the research field has gradually expanded from the theoretical framework and norms to the technical level. More scholars in cross-cutting fields are considering how to promote food security from the technical level from the micro-perspective [38]. For example, they use the progress of biotechnology to improve varieties, promote efficient non-toxic pesticides, develop advanced agricultural technologies, improve water conservancy irrigation systems, establish agricultural information systems, and strengthen the application of information technology in agricultural management to achieve food security.

4.2. *Sustainable Development Path of Food Security*

The goal of food security research is to achieve food security in quantity, quality, and nutrition. Although the grain output still maintains a steady upward trend, and the nutritional level continues to improve, guaranteeing food security still faces enormous challenges, and the food security situation is still severe. Climate change poses many food security issues. Studies have shown that rising global temperatures and more extreme weather events will reduce the production of major food crops such as wheat, corn, and rice. To adapt to future climate change, adopting improved varieties, changing planting dates, and optimizing irrigation systems are important measures. Considering the productivity and sustainability of food and achieving the dual goals of food security and sustainable development are crucial topics. Agriculture accounts for 34% of greenhouse gas emissions, with the majority of them coming from land-use changes caused by agricultural activities and the rest from the food production supply chain [39]. Achieving the Sustainable Development Goals is important to resolve the contradictions between food production, greenhouse gas emissions, and resource depletion [40,41]. Man Li et al. proposed a comprehensive ecological and economic model to analyze crop phenology and nitrogen fertilizer absorption capacity, indicating that the yield can be improved by improving the efficiency of nitrogen fertilizer use [42].

Agro-ecosystems are heavily dependent on water resources, and an increase in food demand will inevitably lead to an increase in water demand [43]. Many scholars evaluated the importance of water resources security to food security [44,45]. Determining how to achieve food security under the premise of ensuring water resources security is an important research topic and involves technology, management, policy, and other aspects [36,46]. Looking at agricultural development from the perspective of sustainable development, it is necessary to explore ecological agriculture models while taking into account the protection of the ecological environment. For example, Lucantoni et al. analyzed the transition to eco-agriculture on a farm in Cuba [47]. Biodiversity conservation has been largely investigated by the international community, with an emphasis on balancing food security and biodiversity conservation. For example, JZA B et al. identified 10 hot-spots that would face food insecurity and biodiversity loss and called for attention to the status quo of conflict hot-spots in an effort to mitigate conflict [48,49]. Faced with the current situation that terrestrial grain productivity is constrained by future arable land potential and water resource depletion, Christopher et al. proposed that expanding the marine grain production capacity through timely fishery reforms would provide a guarantee for food security [50].

4.3. Future Research Direction

According to our analysis of the existing research results, we believe that the following issues in the field of food security will receive attention in the future:

(1) Improvement of the evaluation system for the food security level: The global food security measurement requires a complete indicator system. Demographic, climate, financial (agricultural subsidies, trade restrictions), and policy factors combined influence food security. In 2000, the Committee on World Food Security comprehensively considered consumption, health, and nutrition and formed seven monitoring indicators: (a) the incidence of undernourished population; (b) per capita dietary energy supply; (c) the proportion of grain and rhizome food calories in the per capita dietary energy supply; (d) life expectancy at birth; (e) mortality rate of children under 5 years; (f) proportion of underweight children under 5 years; (g) proportion of adults with a body mass index >18.5. Although the developed countries have sufficient access to food as a whole, local food insecurity is the main problem. For many developing countries, the root cause of food security is still food insufficiency caused by poverty. Numerous publications related to the indicator system have also been produced at national and regional levels, but the current evaluation indicator system is still not comprehensive enough and can only reflect part of the food security level. Food security at the micro-level has gradually become a hot topic, and the measures at the family and individual levels need to be further improved.

(2) Balancing sustainable development and food security: In the early stage, the degradation of natural resources, the decline in farmland production potential [51], and water shortage seriously threatened food security. Many studies focused on agricultural sustainability [52], and the determination of how to ensure food security under the premise of sustainable development is an important research direction at this stage.

(3) Research on improving agricultural production and management efficiency: There are still defects in grain production, sales, and management, along with an unreasonable allocation of resources and an untimely market supply. Improving the quality of agricultural products [53], adjusting the structure of agricultural production [54], and enhancing the popularity of large-scale mechanized agricultural production [55] are all methods to improve agricultural production efficiency. For example, PK Adom et al. showed that technical efficiency in Africa seriously restricted the potential of agricultural output and that technical support can largely contribute to food production in this region [56].

(4) Research on government food security policies and strategies: Currently, there are more severe food shortages in the world's hunger hot-spots. In the context of intensive agricultural development, the rapid development of high-tech fields and the formation of technical barriers have widened the gap in the level of food security among countries. In the case of insufficient policy and institutional guarantees, along with the development of technology, the agricultural technology gap between developed and undeveloped countries will widen increasingly, making the poor population face a more serious food crisis. In the face of conflicts and disputes, natural disasters, and public health emergencies, the type of food security policies should also be the focus of sustainable food security research.

4.4. Comparative Analysis with Relevant Studies

The comparative analysis of relevant studies helps to determine differences, make up for deficiencies, and enrich research results. At present, research on food security involves factors such as concept, population, production, circulation, trade, policy, and the environment. The research perspective is multi-dimensional, and it can deeply discuss some specific issues. The literature is fruitful, and this research field is receiving increasing attention. However, the discussions on food security are scattered, ignoring the systematic characteristics of the research object, and it is difficult to form a unified logical main line connecting different research perspectives. We understood the development of food security research from a macro-perspective and noted the correlation between major research groups and countries as well as the distribution of research topics; we also analyzed the research results of major articles and authors. Our research attempts to clarify the cognitive context of academic circles on hot issues of food security help scholars understand the latest content of food security and can provide a reference for in-depth research on food security issues.

4.5. Limitations of the Study

This paper only used the core dataset in the Web of Science database as the data source. According to the retrieval conditions, the publications were from 1991 to 2022. Undeniably, there were some important research achievements before 1991. In addition, to exclude the interference of non-whole years with the research results, our research includes studies published before 31 December 2021. Therefore, we only analyzed the research results related to food security in the past 30 years. Our research is based on a single database, ignoring articles included in Scopus, CNKI, and other databases. In our following study, we will therefore explore how to incorporate other databases according to the current discussion, with the aim of expanding our research horizon.

5. Conclusions

In this study, we quantitatively analyzed the current state of research in the field of food security, based on the literature related to food security in the Web of Science database and combining the advantages of different bibliometric software packages. The results show that in the past 30 years, food security has experienced three stages of development, namely the initial stage, the development stage, and the high-yield and active stage; scholars' understanding of the connotation of food security has been enriched and improved; and the research results have gradually changed from theory to application and are more practical and operable. The journals *Food Security*, *Sustainability*, and *Food Policy* published the largest numbers of articles and are dedicated to the discussion of food security issues, focusing on the publication of comprehensive views from interdisciplinary perspectives. Food security research covers 138 countries or regions, and the United States is at the vanguard in both quantity and quality. China is the developing country with the most significant contribution to food security research. Food security in sub-Saharan Africa, India, and other regions has received extensive attention. It is worth noting that most mature countries and institutions studying food security are located in Europe. Therefore, strengthening international cooperation in food security is the key to solving the world food problem. Climate change is the most frequently occurring keyword in the field of food security, followed by agriculture and management. The food security theme mapping shows that food production, climate change, and sustainable development had the highest research centrality degree and densities in the field of food security. Poverty in sub-Saharan Africa and India is an important theme for food security. At the same time, it is predicted that ecological agriculture, nutritional security, and food sovereignty will be research topics with good prospects.

To realize the world food security vision, the increasingly complex context of the global economy, including global climate change, production factors, technological progress, and international cooperation, needs to be considered. Maintaining and developing food security under the premise of sustainable development is the core of food security. Improving the food security index system, balancing the relationship between sustainable development and food security, improving the efficiency of agricultural production and management, and research on government food security policies and strategies will be the focus of the future.

Author Contributions: J.L.: methodology, software, data curation, writing—original draft; W.S.: conceptualization, funding acquisition, writing—review and editing. All authors have read and agreed to the published version of the manuscript.

Funding: This research was supported by the Second Tibetan Plateau Scientific Expedition and Research (Grant No. 2019QZKK0603) and the Project of National Natural Science Foundation of China (Grant No. 42071233).

Data Availability Statement: Not applicable.

Conflicts of Interest: The authors declare no conflict of interest. The funders had no role in the design of the study; in the collection, analyses, or interpretation of data; in the writing of the manuscript; or in the decision to publish the results.

References

1. Transforming Our World: The 2030 Agenda for Sustainable Development. Available online: https://sdgs.un.org/2030agenda (accessed on 11 July 2021).
2. Heggen, K.; Sandset, T.J.; Engebretsen, E. COVID-19 and sustainable development goals. *Bull. World Health Organ.* **2020**, *98*, 646. [CrossRef] [PubMed]
3. FAO; IFAD; UNICEF; WFP; WHO. The state of food security and nutrition in the world 2021. In *Transforming Food Systems for Food Security, Improved Nutrition and Affordable Healthy Diets for All*; FAO: Rome, Italy, 2021.
4. Fujimori, S.; Hasegawa, T.; Krey, V.; Riahi, K.; Bertram, C.; Bodirsky, B.L.; Bosetti, V.; Callen, J.; Després, J.; Doelman, J. A multi-model assessment of food security implications of climate change mitigation. *Nat. Sustain.* **2019**, *2*, 386–396. [CrossRef]
5. Ferreira Filho, S.; Horridge, M. Land use change, ethanol production expansion and food security in Brazil. In *Handbook of Bioenergy Economics and Policy*; Khanna, M., Scheffran, J., Zilberman, D., Eds.; Springer: New York, NY, USA, 2010; Volume 2, pp. 303–320. [CrossRef]
6. Guo, Z.; Hu, Y.; Zheng, X. Evaluating the effectiveness of land use master plans in built-up land management: A case study of the Jinan Municipality, eastern China. *Land Use Policy* **2020**, *91*, 104369. [CrossRef]
7. Foley, J.A.; Ramankutty, N.; Brauman, K.A.; Cassidy, E.S.; Gerber, J.S.; Johnston, M.; Mueller, N.D.; O'Connell, C.; Ray, D.K.; West, P.C. Solutions for a cultivated planet. *Nature* **2011**, *478*, 337–342. [CrossRef] [PubMed]
8. Dasgupta, S.; Robinson, E.J. Attributing changes in food insecurity to a changing climate. *Sci. Rep.* **2022**, *12*, 4709. [CrossRef]
9. Song, W.; Liu, M. Assessment of decoupling between rural settlement area and rural population in China. *Land Use Policy* **2014**, *39*, 331–341. [CrossRef]
10. Zou, B.; Mishra, A.K.; Luo, B. Aging population, farm succession, and farmland usage: Evidence from rural China. *Land Use Policy* **2018**, *77*, 437–445. [CrossRef]
11. Mai, Y.; Peng, X.; Chen, W. How fast is the population ageing in China? *Asian Popul. Stud.* **2013**, *9*, 216–239. [CrossRef]
12. Zhang, Y.; Li, X.; Song, W. Determinants of cropland abandonment at the parcel, household and village levels in mountain areas of China: A multi-level analysis. *Land Use Policy* **2014**, *41*, 186–192. [CrossRef]
13. Song, W.; Pijanowski, B.C. The effects of China's cultivated land balance program on potential land productivity at a national scale. *Appl. Geogr.* **2014**, *46*, 158–170. [CrossRef]
14. Zhou, Y.; Liu, Y. Reflections on China's food security and land use policy under rapid urbanization. *Land Use Policy* **2021**, *109*, 105699. [CrossRef]
15. Poonia, V.; Goyal, M.K.; Madramootoo, C.A.; Gupta, A.K.; Saraswat, C. *Impact of COVID-19 on Agro-Food Industry and Transitions Towards Food Security, Integrated Risk of Pandemic: Covid-19 Impacts, Resilience and Recommendation*; Springer: Singapore, 2020; pp. 255–273. [CrossRef]
16. Chanza, N.; Musakwa, W. Revitalizing indigenous ways of maintaining food security in a changing climate: Review of the evidence base from Africa. *Int. J. Clim. Chang. Strateg. Manag.* **2020**, *14*, 252–271. [CrossRef]
17. Arouna, A.; Lokossou, J.C.; Wopereis, M.; Bruce-Oliver, S.; Roy-Macauley, H. Contribution of improved rice varieties to poverty reduction and food security in sub-Saharan Africa. *Glob. Food Secur.* **2017**, *41*, 186–192. [CrossRef]
18. Henriques, P.; O'Dwyer, G.; Dias, P.C.; Barbosa, R.M.S.; Burlandy, L. Health and Food and Nutritional Security Policies: Challenges in controlling childhood obesity. *Ciência Saúde Coletiva* **2018**, *23*, 4143–4152. [CrossRef] [PubMed]
19. Glnzel, W. Bibliometrics as a research field: A course on theory and application of bibliometric indicators. In *Landolt-Börnstein Group III Condensed Matter*; Springer: Berlin/Heidelberg, Germany, 2003; pp. 6–9. [CrossRef]
20. Fafchamps, M. Cash crop production, food price volatility, and rural market integration in the third world. *Am. J. Agric. Econ.* **1992**, *74*, 90–99. [CrossRef]
21. Pinstrupanderson, P. Food Security: Definition and Measurement. *Food Secur.* **2009**, *1*, 5–7. [CrossRef]
22. Ericksen, P.J. Conceptualizing food systems for global environmental change research. *Glob. Environ. Chang.* **2008**, *18*, 234–245. [CrossRef]
23. Carletto, C.; Zezza, A.; Banerjee, R. Towards better measurement of household food security: Harmonizing indicators and the role of household surveys. *Glob. Food Secur.* **2013**, *2*, 30–40. [CrossRef]
24. Cordell, D.; Drangert, J.-O.; White, S. The story of phosphorus: Global food security and food for thought. *Glob. Environ. Chang.* **2009**, *19*, 292–305. [CrossRef]
25. Headey, D.; Ecker, O. Rethinking the measurement of food security: From first principles to best practice. *Food Secur.* **2013**, *5*, 327–343. [CrossRef]
26. Mirzabaev, A.; Olsson, L.; Kerr, R.B.; Pradhan, P.; Lotze-Campen, H. Climate Change and Food Systems. *Annu. Rev. Environ. Resour.* **2021**, *37*, 195–222. [CrossRef]

27. Kummu, M.; Moel, H.D.; Porkka, M.; Siebert, S.; Varis, O.; Ward, P.J. Lost food, wasted resources: Global food supply chain losses and their impacts on freshwater, cropland, and fertiliser use. *Sci. Total Environ.* **2012**, *438*, 477–489. [CrossRef] [PubMed]
28. Maxwell, S. Food security: A post-modern perspective. *Food Policy* **1996**, *21*, 155–170. [CrossRef]
29. Mu, J.; Khan, S. The effect of climate change on the water and food nexus in China. *Food Secur.* **2009**, *1*, 413–430. [CrossRef]
30. Lu, Y.; Zhang, Y.; Hong, Y.; He, L.; Chen, Y. Experiences and lessons from Agri-Food system transformation for sustainable food security: A review of China's practices. *Foods* **2022**, *11*, 137. [CrossRef] [PubMed]
31. Du, J.; King, C. Unravelling China's food security puzzle, 1979–2008. *China Q.* **2018**, *235*, 804–827. [CrossRef]
32. Jones, A.D.; Ngure, F.M.; Pelto, G.; Young, S.L. What are we assessing when we measure food security? A compendium and review of current metrics 1,2. *Adv. Nutr.* **2017**, *4*, 481–505. [CrossRef]
33. Grzelak, A.; Sapa, A. Food Security Problems in Sub-Saharan African Countries. In Proceedings of the Proceedings of the International Scientific Conference "Economic Sciences for Agribusiness and Rural Economy", Warsaw, Poland, 7–8 June 2018. [CrossRef]
34. Legwegoh, A.F.; Fraser, E.D. Food crisis or chronic poverty: Metanarratives of food insecurity in Sub-Saharan Africa. *J. Hunger. Environ. Nutr.* **2015**, *10*, 313–342. [CrossRef]
35. Callon, M.; Courtial, J.P.; Laville, F. Co-word analysis as a tool for describing the network of interactions between basic and technological research: The case of polymer chemsitry. *Scientometrics* **1991**, *22*, 155–205. [CrossRef]
36. Jones, A.D.; Shapiro, L.F.; Wilson, M.L. Assessing the Potential and Limitations of Leveraging Food Sovereignty to Improve Human Health. *Front. Public Health* **2015**, *3*, 263. [CrossRef]
37. Coates, J. Build it back better: Deconstructing food security for improved measurement and action. *Glob. Food Secur.* **2013**, *2*, 188–194. [CrossRef]
38. Wangu, J. The need for a food systems approach in smallholder food and nutrition security initiatives: Lessons from inclusive agribusiness in smallholder communities. *Foods* **2021**, *10*, 1785. [CrossRef] [PubMed]
39. Aggarwal, P.; Vyas, S.; Thornton, P.; Campbell, B.M. How much does climate change add to the challenge of feeding the planet this century? *Environ. Res. Lett.* **2019**, *14*, 043001. [CrossRef]
40. Smith, P.; Gregory, P.J. Climate change and sustainable food production. *Proc. Nutr. Soc.* **2013**, *72*, 21–28. [CrossRef] [PubMed]
41. Shen, X.; Zhang, L.; Zhang, J. Ratoon rice production in central China: Environmental sustainability and food production. *Sci. Total Environ.* **2020**, *764*, 142850. [CrossRef]
42. Li, M.; Guo, Z.; Zhang, W. Balancing food security and environmental sustainability by optimizing seasonal-spatial crop production in Bangladesh. *Environ. Res. Lett.* **2021**, *16*, 074046. [CrossRef]
43. Mekonnen, M.M.; Gerbensleenes, W. The Water Footprint of Global Food Production. *Water* **2020**, *12*, 2696. [CrossRef]
44. McNeill, K.; Macdonald, K.; Singh, A.; Binns, A.D. Food and water security: Analysis of integrated modeling platforms. *Agric. Water Manag.* **2017**, *194*, 100–112. [CrossRef]
45. Cansino-Loeza, B.; del Carmen Munguía-López, A.; Ponce-Ortega, J.M. A water-energy-food security nexus framework based on optimal resource allocation. *Environ. Sci. Policy* **2022**, *133*, 1–16. [CrossRef]
46. Loë, R.; Bjornlund, H. *Governance and the Challenges of Water and Food Security*; WIT Press: Southampton, UK, 2010; Volume 37, pp. 43–55. [CrossRef]
47. Lucantoni, D. Transition to agroecology for improved food security and better living conditions: Case study from a family farm in Pinar del Río, Cuba. *Agroecol. Sustain. Food Syst.* **2020**, *44*, 1124–1161. [CrossRef]
48. Shetty, P. From food security to food and nutrition security: Role of agriculture and farming systems for nutrition. *Curr. Sci.* **2015**, *109*, 456–461. [CrossRef]
49. Jza, B.; Yue, C.; Le, Y.; Xle, F.; Rui, Y.; Peng, G. Future global conflict risk hotspots between biodiversity conservation and food security: 10 countries and 7 Biodiversity Hotspots. *Glob. Ecol. Conserv.* **2022**, *34*, e02036. [CrossRef]
50. Free, C.M.; Cabral, R.B.; Froehlich, H.E.; Battista, W.; Ojea, E.; O'Reilly, E.; Palardy, J.E.; García Molinos, J.; Siegel, K.J.; Arnason, R. Expanding Ocean food production under climate change. *Nature* **2022**, *605*, 490–496. [CrossRef] [PubMed]
51. Song, W. Land-use/land-cover change and ecosystem service provision in China. *Sci. Total Environ.* **2017**, *576*, 705–719. [CrossRef] [PubMed]
52. PérezEscamilla, R. Food Security and the 2015-2030 Sustainable Development Goals: From Human to Planetary Health: Perspectives and Opinions. *Curr. Dev. Nutr.* **2017**, *1*, e000513. [CrossRef] [PubMed]
53. Chaiken, C. Intensive Farming, Agro-Diversity, and Food Security Under Conditions of Extreme Population Pressure in Western Kenya. *Hum. Ecol.* **2000**, *28*, 19–51. [CrossRef]
54. Lu, S.S.; Liu, Y.S.; Long, H.L.; Xing-Hang, G. Agricultural Production Structure Optimization: A Case Study of Major Grain Producing Areas, China. *J. Integr. Agric.* **2013**, *12*, 184–197. [CrossRef]
55. Oyetunde-Usman, Z.; Olagunju, K.O. Determinants of Food Security and Technical Efficiency among Agricultural Households in Nigeria. *Economies* **2019**, *7*, 103. [CrossRef]
56. Adams, S.; Coomes, O.T. Decomposition of technical efficiency in agricultural production in Africa into transient and persistent technical efficiency under heterogeneous technologies. *World Dev.* **2020**, *129*, 104907. [CrossRef]

foods

MDPI

Review

The Role of Alternative Crops in an Upcoming Global Food Crisis: A Concise Review

Antonios Mavroeidis, Ioannis Roussis and Ioanna Kakabouki *

Laboratory of Agronomy, Department of Crop Science, Agricultural University of Athens, 11855 Athens, Greece
* Correspondence: i.kakabouki@gmail.com

Abstract: Achieving Food Security (FS) is perhaps our most challenging aspiration. Despite our best efforts, millions of people around the globe are malnourished or live with hunger. The state of the geo-political scene, as well as the COVID-19 pandemic, have recently brought forth fears of a Global Food Crisis (GFC). Here, we present the factors that threaten FS and could trigger a GFC, examine the potential of alternative crops (ACs) as a measure against an upcoming GFC, and highlight the key aspects of the ACs introduction process in new regions. ACs could enhance FS, yet their success is premised on the adoption of sustainable practices and the implementation of food strategies that aim to promote healthy consumer behaviours.

Keywords: alternative crops; Global Food Crisis; Food Security; Sustainable Development Goals

1. What Makes a Food Crisis?

Since the dawn of mankind, hunger has been our omnipotent enemy. From simple food shortages to famines, whether on a national, regional, or global scale, human history is filled with hunger. By the late 1800s, Walford had recorded more than 70 famines during the 18th–19th centuries throughout Europe, Asia, Africa, and South America [1]. Following the Green revolution of the 1950s–1960s, the frequency, intensity, and mortality of famines were drastically reduced [2]. Nevertheless, malnutrition still affects at least 200 million people [3]. Nowadays, the term "food crisis" is frequently being used when discussing the malnutrition conundrum. Based on the definition by Timmer [4], a food crisis is "the sharp rise of hunger and malnutrition rates at local, national, or global levels". Recently, mainly due to the COVID-19 pandemic and the invasion of Ukraine, fears of an upcoming global food crisis (GFC) have arisen [5,6].

Any food crisis could be perceived as a complicated nexus of socioeconomic and environmental factors. Broadly speaking, any phenomenon that threatens food security (FS) could constitute a driving force behind a food crisis (i.e., armed conflicts, economic busts, and climate change). Being a major threat to agriculture, climate change has increased food insecurity globally. In a study by Dasgupta and Robinson [7], the authors concluded that for every 1 °C of temperature increase, severe global food insecurity was raised by 1.64% during 2019. The findings of this study also suggested that countries with weak economies are particularly susceptible to climate change-induced food insecurity [8]. Besides the increasing temperature, droughts are expected to become more frequent and more intense [9], and soil salinization could be exacerbated, especially in dryland areas [10]. Climate shifts might increase insect and pathogen pressure and reduce pesticide efficacy [10,11].

Climate change mitigation would decrease the chances of a food crisis. However, designing an effective mitigation policy could be tricky. According to Hasegawa et al. [12], the implementation of a horizontal and strict mitigation policy could destabilize the prices and the supply chains of key agricultural commodities, and thus dent global food security. Such destabilizations are known to boost malnutrition rates. In fact, the relatively recent

GFC threat we faced during 2007–2008 was heavily attributed to the elevated prices of basic food products [4,13]. Disturbingly, food prices have once again skyrocketed. The update on the World Food Situation, released by the FAO on 8 April 2022, reported an all-time high record of the Food Price, Cereal Price, Vegetable Oil Price, and Meat Price Indices [14]. The incremental tendency of these FAO indices was anticipated following the invasion of Ukraine. Nonetheless, prior to the invasion, or even the outbreak of COVID-19, the existing climate scenarios predicted that the prices of wheat, maize, and rice would increase by more than 30% by 2050 [15]. The pandemic only deteriorated the situation, as it caused perhaps the most severe post-WWII economic downturn [6].

Concurrently, the Russo-Ukrainian conflict is considered by many a food-security ticking bomb [16]. War is a major, if not the main, driver that pushes people to malnutrition and hunger. According to the latest report on GFC by the Global Network Against Food Crises (GNAFC), during 2021, armed conflicts drove nearly 140 million people from 24 countries/territories into food crisis [3]. However, the invasion of Ukraine is more than an armed conflict. Both the Russian Federation and Ukraine are major net exporters of agricultural products, and leading suppliers of agro-food commodities in the global markets [17] (Figure 1). Notably, in 2021 these two countries were amongst the top global exporters of wheat, maize, rapeseed, sunflower seeds, and sunflower oil [17]. In the same year, the Russian Federation was the top, second, and third largest exporter of nitrogen, potassium, and phosphorus fertilizers, respectively [17]. The conflict between these two countries could disrupt the global food, fertilizer, and fuel systems and supply chains, plunging millions into hunger [16].

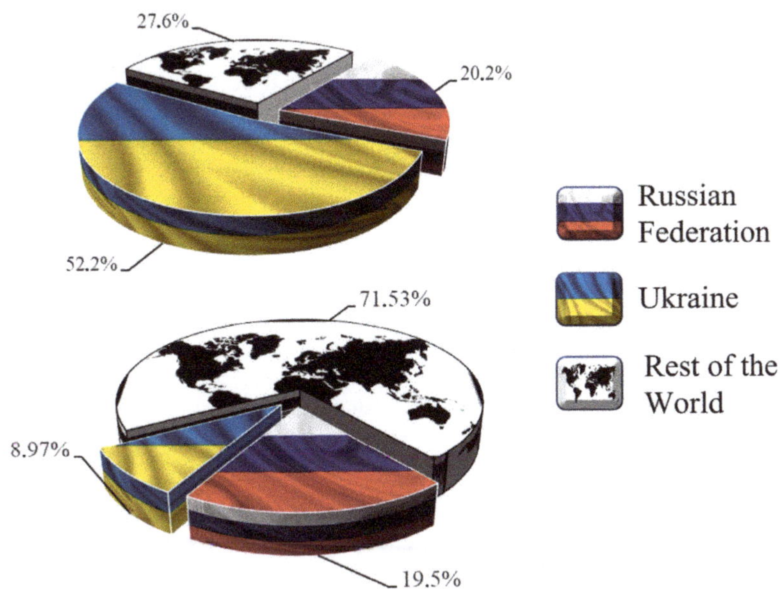

Figure 1. The share of Ukraine, the Russian Federation, and the rest of the world in the global exports of sunflower-seed or oil (crude) on the top, and wheat on the bottom, during 2021. Data obtained from the official website of the Observatory of Economic Complexity (OEC) (https://oec.world/en) (accessed on 14 October 2022).

On top of all that, the world is amidst the COVID-19 pandemic, which impaired agricultural production [6] and economic activities [18]. As lockdowns came into force all around the globe, transportation restrictions led to manpower scarcity on a farm level [6]. This was also the case for the food-processing industry, as gradually more and more workers contracted COVID-19 and food-industry plants were forced to temporarily halt production or operate at much slower rates [19]. Food trade restrictions negatively affected food distribution [19]. Once again, countries with weaker economies, poor healthcare systems, and labor-intensive agricultural sectors were found to be less resilient to the COVID-19-induced shocks to agriculture and to food-supply chain disruptions [6,19].

Some might argue that malnutrition mainly affects the developing countries and that the countries of the first world are not in an immediate threat of a food crisis. For instance, despite the aforementioned information, the European Union (EU) is self-sufficient in the majority of key agricultural commodities, and the availability of food within the EU is probably not at risk [20]. However, the accessibility of food prices, as well as the availability of fertilizers and animal feed, are still open to doubt [20]. Based on the above, the status quo is at least alarming as the threat of a GFC is lurking in the horizon. It is possible that we are already in the midst of such a crisis. In the 2022 Global Report on Food Crisis, the Secretary-General of the United Nations stated that "we are facing hunger on an unprecedented scale, food prices have never been higher, and millions of lives and livelihoods are hanging in the balance" [3]. Of course, managing such a threat is a convoluted task that requires coordinated interdisciplinary collaborations. Here, we will focus on the potential, strengths, and weaknesses of alternative crops.

2. Alternative Crops and Food Security

Initially, we need to define what an alternative (or novel, innovative, retrovative, etc.) crop is. In most cases, the alternative crops (ACs) are described as crops that can be introduced into a new agroecosystem in lieu of "traditional" crops that are usually more susceptible to biotic (e.g., pests) and abiotic stress [21] (e.g., salinity). For instance, heritage cereals that are usually more resilient to extreme weather events compared to the modern cereal varieties [22,23], could be characterized as ACs. According to Isleib [24], ACs are crops (re)introduced in a particular geographic area due to their potential high value or other benefits to the farming systems of that area. ACs are frequently mixed up with the underutilized species (NUS) (also known as neglected, orphan, or niche crops) [25]. NUS were primarily being cultivated in their center of origin; however, at some point in time they lost favor and now have regained interest (locally or in a wider scale) [26]. The concepts of ACs and NUS have apparent differences, yet some NUS could be perceived as ACs, provided they have been proposed as promising crops to be (re)introduced to an area of adaptation. Subsequently, there is no strict classification of a group of crops as ACs (Table 1). As a case in point, teff is regarded as an AC in the Mediterranean Basin [27], yet as a traditional crop in Ethiopia [28].

In order to evaluate the potential role of ACs in an upcoming GFC, initially we have to examine their beneficial effects on food security (FS). According to the definition by FAO, FS is the state when "all people have physical and economic access to sufficient, safe, and nutritious food that meets their dietary needs and food preferences for an active and healthy life" [29]. Of course, "access" is not the only factor that defines FS. FS is founded on four pillars: availability, access, utilization, and stability.

The availability of food is both a quantitative, and a qualitative indicator, as it refers to the existence of sufficient amounts of domestically produced and/or imported nutritious food [30]. To enhance food availability, the introduction of an AC to a region should aim to increase the quantities of produced food commodities in that region, and offer high-quality, nutritious food alternatives. Several crops rich in micro- and macronutrients have been proposed as ACs (Table 1). The literature also highlights the acclimatization and adaptation potential of some of them under high salinity, drought conditions, water logging, and in soils with low fertility [26,27]. The introduction of ACs to such low-productivity areas could increase food availability.

Table 1. List of crops that have been proposed as ACs [21,27], their common and scientific names, family, area of origin, and nutritional value. These crops have been proposed as ACs due to their acclimatization potential to marginal environments and/or their tolerance to biotic (pests and diseases) or abiotic stress (high salinity and sodicity, droughts, and high temperatures).

Common Name	Scientific Name	Family	Area of Origin	Nutritional Value	Referance
Amaranthus	Amaranthus retroflexus	Amaranthaceae	Americas	Seeds abundant in protein content (13–19%), high levels of oils rich in squalene, and high amounts of antioxidants	[31]
Buckwheat	Fagopyrum esculentum	Polygonaceae	Asia	Protein content similar to that of wheat, aproximantely 3% fat content, and high crude fiber concentration	[32]
Canihua	Chenopodium pallidicaule	Amaranthaceae	Andes	Exeptional protein, fat, ash, and carbohydrate content	[33]
Einkorn	Triticum monococcum	Poaceae	Asia Minor	Rich in antioxidant compounds such ascarotenoids, tocols, conjugated polyphenols, alkylresorcinols, and phytosterols	[34]
Emmer wheat	Triticum dicoccon	Poaceae	Eurasia	Rich in resistant starch, minerals, fibre, carotenoids, and antioxidant compounds	[35]
Foxtail	Setaria italica	Poaceae	Southern Asia	Rich in protein, fatty acids, minerals, and amino acids	[36]
Khorosan wheat	Triticum turgidum ssp. Turanicum	Poaceae	Mesopotamia	Higher protein, crude ash, and vitamine E content compared to wheat	[37]
Pearl millet	Cenchrus americanus	Poaceae	West Africa	360 calories, 12 g of protein, 5 g of fat, 1 g of fibres, and 67 g of carbohydrates per 100 g of seeds	[38]
Quinoa	Chenopodium quinoa	Amaranthaceae	Andes	Gluten-free, with high protein concent, rich in unsaturated fatty acids, vitamins, and minerals	[39]
Salicornia	Salicornia bigelovii	Amaranthaceae	North America	Rich in bioactive compounds, vitamin A, minerals and fatty acids. Seedoil rich in linoleic acid	[40]
Spelt	Triticum spelta	Poaceae	Europe	Higher protein content, more non-essential amino acids, and less lysine	[41]
Tef	Eragrostis tef	Poaceae	Somali Peninsula	Gluten-free, rich in protein, dietary fiber, polyphenols, and minerals	[42]
Triticale	×Triticosecale	Poaceae	Europe	High protein content and slightly higher levels of most of the nutritious compounds when compared to wheat	[43]
Tritordeum	Tritordeum martinii	Poaceae	Europe	High total phenol content, antioxidant activity, dietary fiber content, and total free amino acids	[44]
Cowpea	Vigna unguiculata	Fabaceae	Southern Africa	Rich in protein (<20%) and minerals (calcium, potassium, sodium, and more)	[45]

Table 1. *Cont.*

Common Name	Scientific Name	Family	Area of Origin	Nutritional Value	Referance
Guar	Cyamopsis tetragonolobus	Fabaceae	Africa	High protein, ash, and polyphenol contents	[46]
Lablab	Lablab purpureus	Fabaceae	South-east Asia	Rich in proteins, carbohydrates, minerals and vitamins	[47]
White lupin	Lupinus albus	Fabaceae	Mediteranean Basin	Fair protein, fatty acid, and fibre content, as well as oligosaccharides, antioxidants, and non-starch carbohydrates	[48]
Pigeon pea	Cajanus cajan	Fabaceae	South Asia	Rich in starch, protein, calcium, manganese, crude fibre, fat, and minerals	[49]
Sesbania	Sesbania sp.	Fabaceae	North-East Africa	High protein content (can exced 40%), vitamin C, and calcium	[50]
Indian mustard	Brassica juncea	Brassicaceae	West Asia	Seeds rich in glucosinolates, sterols, and phenols. Leafs rich in glucose, fructose, and minerals	[51,52]
Purslane	Portulaca oleracea	Portulacaceae	Eurasia	Rich in omega-3, amino acids, and vitamins	[53]
Chia	Salvia hispanica	Lamiaceae	Central America	Seeds with high protein content (>15%), rich in lipids, and minerals. On average, 100 g of seed contains aproximantely 500 kcal	[54]
Nigella	Nigella sativa	Ranunculaceae	Eastern Europe	Rich in fatty acids, phytosterols, glycolipids, and phospholipids	[55]
Sweet potato	Ipomoea batatas	Convolvulaceae	Americas	Protein content ranging from 4–27%, rich in β-carotene and anthocyanin	[56]
Camelina	Camelina sativa	Brassicaceae	Europe	Excelent source of essential unsaturated fatty acids, particularly OMEGA-3 fatty acids	[57]

As marginal areas constitute a significant portion of available land, their exploitation via the introduction of stress tolerant ACs would increase their agricultural productivity, and thus enhance FS [21]. For instance, in Iran, high salinity and droughts are major obstacles to agriculture [58]. The experimental incorporation of quinoa, a salinity tolerant crop, in Iran reported promising results as irrigation with 14 dS m^{-1} of saline water resulted in grain yields of 2–3 t/ha [21].

Access to food is mainly determined by economic factors [59]. Typically, the balance between food prices and household income/assets influences food access. The prices of agricultural commodities usually depend on their supply and demand [60]. Shifts in their supply/demand equilibrium tend to alter their price. This affects FS, as extreme downward or upward price oscillations have been proven to be detrimental to food access [61]. From a financial point of view, ACs can boost food access to rural areas by improving the agriculture household income, especially in the case of small- and medium-scale farmers [62]. Moreover, ACs promote food/crop diversity. Crop diversification can stabilize the farmers' flow of income, especially in a small-farm scale [63]. Additionally, they provide resilience to income shocks and opportunities for improved incomes, due to the potentially favorable prices of ACs [64–66].

The FAO defines food utilization as "the proper biological use of food", under the context of a healthy diet that provides sufficient energy and essential nutrients [67]. Modern agriculture relays heavily on a few staple food crops to meet the global demand. It is estimated that wheat, maize, and rice provide more than half of the world's plant-derived calories [26]. That being the case, a lack in dietary variety can lead to malnutrition despite adequate caloric intake [59]. Enriching crop diversity would benefit both the FS and the natural agroecosystems [68]. The introduction of ACs could increase the versatility and improve the nutritional content of meal plans, as the grains of several ACs are rich in proteins, fats, crude fibers, micro- and macronutrients, etc. [69]. Moreover, crop diversity is known to benefit nutritional stability [70]. Admittedly, due to their poor market presence, the advancements in processing and storage methods of AC final products are often lackluster [26]. Proper storage is essential for tackling food insecurity [71].

Food stability can be achieved when food availability, access, and utilization are consistent through time [29]. This dimension of FS comprises an expression of the need for sustainable food production and sustainable agricultural systems. Sustainable agriculture itself is compromised of the changing climate, the loss of agricultural biodiversity, soil degradation, and water and air pollution [72]. The intensification of agriculture and the mainstay inputs of conventional agricultural systems only aggravate the situation [73]. On the contrary, the adoption of ACs could tackle these constraints on sustainable agriculture. Besides the enhancement of crop diversity, many ACs often require low (compared to traditional crops) chemical inputs in the form of fertilizers and pesticides [27]. Similarly, the drought tolerant ones are characterized by reduced irrigation needs [27]. Due to their low input needs, they can perform adequately under organic systems [27]. As a result, the introduction and cultivation of ACs could further reduce environmental degradation. In a recent study by Mazac et al. [74], the authors estimated that the incorporation of novel foods in European food systems would contribute to global warming mitigation, as well as improve water and land use by over 80%.

3. Food for Thought

The introduction of ACs should be dealt with caution, otherwise, not only will they not contribute to FS, but they could also be unprofitable for the farmers and damaging to the agricultural systems they are introduced into. To fully understand this dynamic, one can refer to the example of quinoa. Quinoa originates from the South American Andes, where it has been cultivated for more than 8000 years [75]. Following the 1950s, quinoa gained international attention, due to its high nutritional value, that peaked around the mid-2010s [76]. As the demand for quinoa grew, the crop was introduced to many countries in Europe and North America, though they continued to import significant amounts of quinoa from the three major producers (Peru, Bolivia, and Ecuador) [75,76]. Quinoa markets boomed, the demand increasing rapidly, and the international prices were elevated. In the Andes, the attractive prices of quinoa shifted its cultivation from small-farm "traditional" models to large-scale, market-oriented farming [75]. Cultivation of quinoa was intensified to the point that land use changes, land degradation, extensive monocropping, and the loss of genetic diversity threatened the sustainability of both the production of quinoa and the local agroecosystems [75,76]. Soon, the production of quinoa in the Andes doubled, the supply of quinoa exceeded the global demand, and the prices fell [76]. This phenomenon is known as "boom and bust", a chain reaction catalyzed by market trends, that results in acute price oscillations and a shift towards less sustainable agricultural practices [76].

Andreotti et al. [76] acknowledged that the "boom and bust" of such crops can be divided in stages: promotion, boom, bust, and transition to a new system. These stages could function as the pillars upon which policymakers can design frameworks for the introduction of ACs to new areas. Here, we will highlight key features of the ACs integration process that, based on the literature, as well as empirical knowledge, should always be regarded. Based on the work of Andreotti et al. [76], this process could be divided into three phases: promotion of ACs, incorporation to food systems, and sustainable production, and they should aim to avoid the "boom and bust" phenomenon (Figure 2).

Promotion of ACs: Promoting an AC is based on the simple, yet admittedly challenging task of raising awareness and educating the public. This can be done via mainstream and social media, workshops and living labs, educational initiatives, or even peer-to-peer interactions. National governments possess the means (e.g., taxations and subsidies) to motivate farmers to adopt ACs. Farmers on their behalf should comprehend every beneficial aspect of the ACs on FS and sustainable agricultural practices, instead of focusing solely on their potential short-term economic returns. This tendency has also been observed on the adoption rates of integrated pest management strategies [77]. However, well-informed farmers have been reportedly more likely to change their attitudes towards these practices [78]. Well-informed farmers might also be more likely to adopt ACs. Consumers, on the other hand, might already be more willing to embrace ACs. A recent study by Wendin et al. [22] found that, in the case of heritage cereals, women and elders amongst the different age/sex groups were the most concerned regarding the origin and health benefits of the AC, and the elderly were more willing to pay higher prices for the AC products. They also reported that in the majority of the participants in their study were aware of the heritage cereals (to least at some extent). The authors attributed this finding partially to the recent health trends that have been related with such ACs [22]. However, consumers (mainly in developed countries) should adopt proper attitudes and not simply follow food trends.

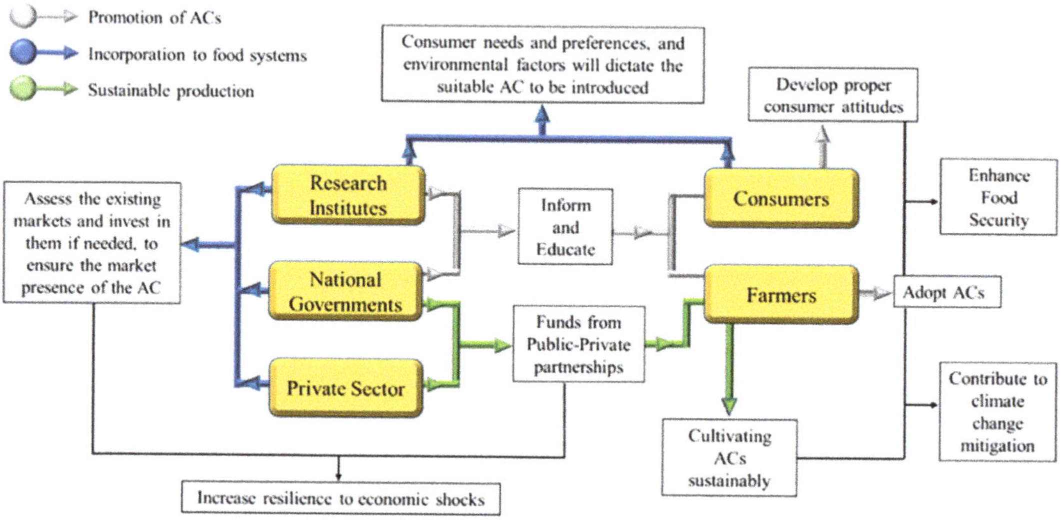

Figure 2. Highlights of the three stages of introducing ACs. All the actors (yellow color) of the ACs value chain and their key interactions are briefly depicted in the figure. The actions and interactions of each stage are depicted with different colors (white, blue, and green). As the three stages are not necessarily successive, actions and interactions of the actors from different stages could be simultaneous.

At this point, a paradox needs to be addressed. As mentioned above, nearly 200 million people are currently in food crisis all over the world [3]. Yet more than 700 million are on the opposite side of the nutritional spectrum, being overweight or obese [79]. The rates of obesity are expected to increase, and by 2030, more than 1 billion people are estimated to be living with obesity [79]. Obesity is rapidly turning into an epidemic, especially in the developed countries. Recent studies report that more than 40% of North Americans are obese and approximately 60% of Europeans are either overweight or obese [80,81]. The literature strongly suggests that, besides the plethora of health problems that have been attributed to obesity, there is a link between it and COVID-19 high mortality rates [82]. As health experts call for a solution, the timing is perfect for proposing more diverse and healthy diets. ACs could offer viable solutions to meal plans and their promotion could be part of a healthier "new food agenda", that could also target younger audiences (e.g., inclusion of zero food waste and healthy diet-related lectures in school curriculums) and help them develop healthy eating habits from their early years.

Incorporation into food systems: Initially, the selection of the introduced AC is premises on meticulous planning based on region-specific studies. Several factors must be considered, including environmental and pedoclimatic niches, regional food preferences and needs, and the societal benefits of the AC to local communities. The involvement of National Agricultural Research Systems (NARS) is vital. Research should also include breeding programs to improve the crop if needed (e.g., reduce seed heterogeneity). However, the importance of the genetic diversity of the crop should not be neglected, as it relates to the crop's adaptability [75]. The lesson learned from quinoa's boom and bust is that when ACs transition from smallholding to an industrial agriculture model, the crops' genetic variety gets disregarded, due to market pressure [75]. The industrialization of the ACs might also ignore the empirical farming knowledge passing down from generation to generation (especially in the case of NUCs). Ex situ gene banks and the utilization of cultivation-practices related to traditional knowledge will be essential for the improvement of the ACs' performance [26]. Finally, it is crucial to ensure the ACs market presence. This requires the assessment of the dynamics of regional agricultural development, the existing markets and supply chains, and the logistics costs.

Sustainable production: Research on the optimization of cultivation practices (e.g., fertilization, irrigation, etc.) and food processing should be constant and not limited during the introduction of ACs in a new food system. The effects of ACs on the environment and on the everyday lives of rural populations should be regularly evaluated. Governments, civil society organizations, and private partners should monitor the value chains of ACs and interfere when needed to avoid any boom-and-bust scenarios. Overall, the cooperation of both the public and the private sectors (public-private partnerships) would be beneficial for the sustainability of ACs, and the agricultural systems as a whole. A recent report by the UN Food Systems Summit, the World Bank, the International Food Policy Research Institute (IFPRI), and the Food and Land Use Coalition presented the Food Finance Architecture, a five food finance imperatives-based policy for sustainable food systems [83]. Under this context, governments and private sector partners could mutually finance investments with social and environmental impact, such as the incorporation of ACs in food systems.

ACs are very promising for the future of agriculture. After all, the introduction of ACs also complies with the Sustainable Development Goals (SDGs) set by the UN (Figure 3). The sustainability of agriculture has become a major challenge for policy makers all around the globe. Both the EU and the USA, two of the most significant economic regions of the world with vastly contrasting approaches in agriculture, have designed their strategies to achieve that [84]. ACs seem to be fitting for the Special Objectives of the EU's Common Agricultural Policy 2023–2027 [27] (climate change mitigation, biodiversity enhancement, sustainable food production), the aims of the European Green Deal [27] (reduction of chemical inputs, creation of sustainable food labeling, reduction of greenhouse gases emissions), as well as the aspirational goals of the US Agriculture Innovation Strategy [85] (market expansion and diversity). The adoption of ACs could also facilitate the implementation of the Agenda 2063 that aims to enhance Africa's collective FS by 2063 [86], and the 2030 Strategy of the Asian Development Bank that intends to tackle climate change while strengthening FS in Asia by 2030 [87]. ACs can be grown sustainably, but at the same time, they have much to offer to the concept of agricultural sustainability itself.

Figure 3. ACs and the SDGs set by the UN. Each SDG is depicted with a different color and is attributed its corresponding number. The contribution of ACs to the SDGs is depicted in the center of the figure next to the corresponding number of each SDG.

Author Contributions: Conceptualization, A.M. and I.K.; methodology, I.K.; software, I.R.; validation, A.M., I.R. and I.K.; formal analysis, I.R.; investigation, A.M.; resources, I.K.; data curation, I.R.; writing—original draft preparation, A.M.; writing—review and editing, A.M., I.R. and I.K.; visualization, A.M. and I.K.; supervision, I.K.; project administration, I.K. All authors have read and agreed to the published version of the manuscript.

Funding: This research received no external funding.

Institutional Review Board Statement: Not applicable.

Informed Consent Statement: Not applicable.

Data Availability Statement: Not applicable.

Conflicts of Interest: The authors declare no conflict of interest.

Abbreviations

GFC, Global Food Crisis; FS, Food Security; FAO, Food and Agriculture Organization of the United Nations; GNAFC, Global Network Against Food Crises; EU, European Union; ACs, Alternative Crops; NUCs, Neglected and Underutilized Crops; NARS, National Agricultural Research Systems; UN, United Nations; IFPRI, International Food Policy Research Institute; SDGs, Sustainable Development Goals; US, United States.

References

1. Walford, C. The famines of the world: Past and present. *J. Stat. Soc. Lond.* **1878**, *41*, 433–535. [CrossRef]
2. World Peace Foundation. Annual Report, Fiscal Year 2016–2017. Available online: https://sites.tufts.edu/wpf/files/2017/08/World-Peace-Foundation_Annual-Report-FY2017-Website.pdf (accessed on 14 October 2022).
3. WFP-World Food Programme. The Global Report on Food Crises 2022. Available online: https://docs.wfp.org/api/documents/WFP-0000138913/download/?_ga=2.195833519.2005609609.1656526761-878536380.1651662482 (accessed on 14 October 2022).
4. Timmer, C.P. Preventing food crises using a food policy approach. *J. Nutr.* **2010**, *140*, 224–228. [CrossRef] [PubMed]
5. Bentley, A. Broken bread—avert global wheat crisis caused by invasion of Ukraine. *Nature* **2022**, *603*, 551. [CrossRef]
6. Okolie, C.C.; Ogundeji, A.A. Effect of COVID-19 on agricultural production and food security: A scientometric analysis. *Humanit. Soc. Sci. Commun.* **2022**, *9*, 64. [CrossRef]
7. Dasgupta, S.; Robinson, E.J. Attributing changes in food insecurity to a changing climate. *Sci. Rep.* **2022**, *12*, 4709. [CrossRef] [PubMed]
8. Pokhrel, Y.; Felfelani, F.; Satoh, Y.; Boulange, J.; Burek, P.; Gädeke, A.; Gerten, D.; Gosling, N.S.; Grillakis, M.; Gudmundsson, L.; et al. Global terrestrial water storage and drought severity under climate change. *Nat. Clim. Chang.* **2021**, *11*, 226–233. [CrossRef]
9. Hassani, A.; Azapagic, A.; Shokri, N. Global predictions of primary soil salinization under changing climate in the 21st century. *Nat. Commun.* **2021**, *12*, 6663. [CrossRef]
10. Chaloner, T.M.; Gurr, S.J.; Bebber, D.P. Plant pathogen infection risk tracks global crop yields under climate change. *Nat. Clim. Chang.* **2021**, *11*, 710–715. [CrossRef]
11. Ma, C.S.; Zhang, W.; Peng, Y.; Zhao, F.; Chang, X.Q.; Xing, K.; Zhu, L.; Ma, G.; Yang, H.P.; Rudolf, V.H. Climate warming promotes pesticide resistance through expanding overwintering range of a global pest. *Nat. Commun.* **2021**, *12*, 5351. [CrossRef]
12. Hasegawa, T.; Fujimori, S.; Havlík, P.; Valin, H.; Bodirsky, B.L.; Doelman, J.C.; Fellmann, T.; Kyle, P.; Koopman, J.F.L.; Lotze-Campen, H.; et al. Risk of increased food insecurity under stringent global climate change mitigation policy. *Nat. Clim. Chang.* **2018**, *8*, 699–703. [CrossRef]
13. Von Braun, J. The food crisis isn't over. *Nature* **2008**, *456*, 701. [CrossRef] [PubMed]
14. FAO. World Food Situation. Available online: https://www.fao.org/worldfoodsituation/foodpricesindex/en/ (accessed on 14 October 2022).
15. Bioversity International. Agrobiodiversity Index Report 2019: Risk and Resilience. Available online: https://edepot.wur.nl/530472 (accessed on 14 October 2022).
16. Osendarp, S.; Verburg, G.; Bhutta, Z.; Black, R.E.; de Pee, S.; Fabrizio, C.; Headey, D.; Heidkamp, R.; Laborde, D.; Ruel, M.T. Act now before Ukraine war plunges millions into malnutrition. *Nature* **2022**, *604*, 620–624. [CrossRef] [PubMed]
17. FAO. The Importance of Ukraine and the Russian Federation for Global Agricultural Markets and the Risks Associated with the Current Conflict. Available online: https://www.fao.org/3/cb9236en/cb9236en.pdf (accessed on 14 October 2022).
18. Barman, A.; Das, R.; De, P.K. Impact of COVID-19 in food supply chain: Disruptions and recovery strategy. *CRBS* **2021**, *2*, 100017. [CrossRef]
19. Laborde, D.; Martin, W.; Swinnen, J.; Vos, R. COVID-19 risks to global food security. *Science* **2020**, *369*, 500–502. [CrossRef] [PubMed]
20. Russia's war on Ukraine: Impact on Food Security and EU Response. Available online: https://www.europarl.europa.eu/RegData/etudes/ATAG/2022/729367/EPRS_ATA(2022)729367_EN.pdf (accessed on 14 October 2022).
21. Hirich, A.; Choukr-Allah, R.; Ragab, R. *Emerging Research in Alternative Crops*, 1st ed.; Springer International Publishing: Cham, Switzerland, 2020. [CrossRef]
22. Wendin, K.; Mustafa, A.; Ortman, T.; Gerhardt, K. Consumer awareness, attitudes and preferences towards heritage cereals. *Foods* **2020**, *9*, 742. [CrossRef]
23. Slama, A.; Mallek-Malej, E.; Mohamed, H.B.; Rhim, T.; Radhouane, L. A return to the genetic heritage of durum wheat to cope with drought heightened by climate change. *PLoS ONE* **2018**, *13*, e0196873. [CrossRef]
24. Isleib, J. *Exploring Alternative Field Crops*, 1st ed.; Michigan State University Extension: East Lansing, MI, USA, 2012.
25. Padulosi, S.; Phrang, R.; Rosado-May, F.J. *Supporting Nutrition Sensitive Agriculture through Neglected and Underutilized Species: Operational Framework*; IFAD: Rome, Italy, 2019.
26. Padulosi, S.; King, E.I.O.; Hunter, D.; Swaminathan, M.S. *Orphan Crops for Sustainable Food and Nutrition Security: Promoting Neglected and Underutilized Species*, 1st ed.; Routledge: London, UK, 2021.
27. Kakabouki, I.; Tataridas, A.; Mavroeidis, A.; Kousta, A.; Roussis, I.; Katsenios, N.; Efthimiadou, A.; Papastylianou, P. Introduction of alternative crops in the Mediterranean to satisfy EU Green Deal goals. A review. *Agron. Sustain. Dev.* **2021**, *41*, 71. [CrossRef]
28. Girija, A.; Jifar, H.; Jones, C.; Yadav, R.; Doonan, J.; Mur, L.A. Tef: A tiny grain with enormous potential. *Trends Plant Sci.* **2022**, *27*, 220–223. [CrossRef]
29. FAO. Policy Brief-Food security. Available online: https://www.fao.org/fileadmin/templates/faoitaly/documents/pdf/pdf_Food_Security_Cocept_Note.pdf (accessed on 14 October 2022).
30. Gunaratne, M.S.; Radin Firdaus, R.B.; Rathnasooriya, S.I. Climate change and food security in Sri Lanka: Towards food sovereignty. *Humanit. Soc. Sci. Commun.* **2021**, *8*, 229. [CrossRef]
31. Soriano-García, M.; Aguirre-Díaz, I.S. Nutritional functional value and therapeutic utilization of Amaranth. In *Nutritional Value of Amaranth*, 1st ed.; Waisundara, V.Y., Ed.; IntechOpen: London, UK, 2019. [CrossRef]

32. Eggum, B.O.; Kreft, I.; Javornik, B. Chemical composition and protein quality of buckwheat (*Fagopyrum esculentum* Moench). *Plant Foods Hum. Nutr.* **1980**, *30*, 175–179. [CrossRef]
33. Repo-Carrasco, R.; Espinoza, C.; Jacobsen, S.E. Nutritional value and use of the Andean crops quinoa (*Chenopodium quinoa*) and kañiwa (*Chenopodium pallidicaule*). *Food Rev. Int.* **2003**, *19*, 179–189. [CrossRef]
34. Hidalgo, A.; Brandolini, A. Nutritional properties of einkorn wheat (*Triticum monococcum* L.). *J. Sci. Food Agric.* **2014**, *94*, 601–612. [CrossRef] [PubMed]
35. Čurná, V.; Lacko-Bartošová, M. Chemical composition and nutritional value of emmer wheat (*Triticum dicoccon* schrank): A review. *J. Cent. Eur. Agric.* **2017**, *18*, 117–134. [CrossRef]
36. Zhang, B.; Liu, J.; Cheng, L.; Zhang, Y.; Hou, S.; Sun, Z.; Li, H.; Han, Y. Carotenoid composition and expression of biosynthetic genes in yellow and white foxtail millet [*Setaria italica* (L.) Beauv]. *J. Cereal Sci.* **2019**, *85*, 84–90. [CrossRef]
37. Van Boxstael, F.; Aerts, H.; Linssen, S.; Latré, J.; Christiaens, A.; Haesaert, G.; Dierickx, I.; Brusselle, J.; De Keyzer, W. A comparison of the nutritional value of Einkorn, Emmer, Khorasan and modern wheat: Whole grains, processed in bread, and population-level intake implications. *J. Sci. Food Agric.* **2020**, *100*, 4108–4118. [CrossRef]
38. Malik, S. Pearl millet-nutritional value and medicinal uses. *IJARIIE* **2015**, *1*, 414–418.
39. Bastidas, E.G.; Roura, R.; Rizzolo, D.A.D.; Massanés, T.; Gomis, R. Quinoa (Chenopodium quinoa Willd), from nutritional value to potential health benefits: An integrative review. *J. Nutr. Food Sci.* **2016**, *6*, 1000497. [CrossRef]
40. Cárdenas-Pérez, S.; Piernik, A.; Chanona-Pérez, J.J.; Grigore, M.N.; Perea-Flores, M.J. An overview of the emerging trends of the *Salicornia* L. genus as a sustainable crop. *Environ. Exp. Bot.* **2021**, *191*, 104606. [CrossRef]
41. Chrenkova, M.; Čerešňáková, Z.; Sommer, A.; Galova, Z.; Kráľová, V. Assessment of nutritional value in spelt (*Triticum spelta* L.) and winter (*Triticum aestivum* L.) wheat by chemical and biological methods. *Czech J. Anim. Sci.* **2000**, *45*, 133–137.
42. Zhu, F. Chemical composition and food uses of teff (*Eragrostis tef*). *Food Chem.* **2018**, *239*, 402–415. [CrossRef]
43. Biel, W.; Kazimierska, K.; Bashutska, U. Nutritional value of wheat, triticale, barley and oat grains. *Acta Sci. Pol. Zootech.* **2020**, *19*, 19–28. [CrossRef]
44. Arora, K.; Carafa, I.; Fava, F.; Tuohy, K.M.; Nikoloudaki, O.; Gobbetti, M.; Di Cagno, R. Sourdough performances of the golden cereal Tritordeum: Dynamics of microbial ecology, biochemical and nutritional features. *Int. J. Food Microbiol.* **2022**, *374*, 109725. [CrossRef] [PubMed]
45. Khalid, I.I.; Elhardallou, S.B. Factors that compromise the nutritional value of cowpea fluor and its protein isolates. *Food Nutr. Sci.* **2016**, *7*, 112. [CrossRef]
46. Sharma, P.; Kaur, A.; Kaur, S. Nutritional quality of flours from guar bean (*Cyamopsis tetragonoloba*) varieties as affected by different processing methods. *J. Food Sci. Technol.* **2017**, *54*, 1866–1872. [CrossRef]
47. Khan, A.U.; Choudhury, M.A.R.; Talucder, M.S.A.; Hossain, M.; Ali, S.; Akter, T.; Ehsanullah, M. Constraints and solutions of country bean (*Lablab purpureus* L.) Production: A review. *Acta Entomol. Zool.* **2020**, *1*, 37–45. [CrossRef]
48. Janusz, P. White lupin (*Lupinus albus* L.)–nutritional and health values in human nutrition–a review. *Czech J. Food Sci.* **2017**, *35*, 95–105. [CrossRef]
49. Sharma, S.; Agarwal, N.; Verma, P. Pigeon pea (*Cajanus cajan* L.): A hidden treasure of regime nutrition. *J. Funct. Environ. Bot.* **2011**, *1*, 91–101. [CrossRef]
50. Bunma, S.; Balslev, H. A review of the economic botany of Sesbania (Leguminosae). *Bot. Rev.* **2019**, *85*, 185–251. [CrossRef]
51. Vaštakaitė-Kairienė, V.; Brazaityte, A.; Virsile, A.; Samuoliene, G.; Miliauskiene, J.; Jankauskiene, J.; Novickovas, A.; Duchovskis, P. The nutritional value of brassica leafy greens in different growth stages. *Acta Hortic.* **2020**, *1271*, 455–464. [CrossRef]
52. Szőllősi, R. Indian Mustard (*Brassica juncea* L.) Seeds in Health. In *Nuts and Seeds in Health and Disease Prevention*, 1st ed.; Preedy, V.R., Watson, R.R., Patel, V.B., Eds.; Academic Press: Cambridge, MA, USA, 2020; pp. 357–364. [CrossRef]
53. Nemzer, B.; Al-Taher, F.; Abshiru, N. Phytochemical composition and nutritional value of different plant parts in two cultivated and wild purslane (*Portulaca oleracea* L.) genotypes. *Food Chem.* **2020**, *320*, 126621. [CrossRef]
54. Kulczyński, B.; Kobus-Cisowska, J.; Taczanowski, M.; Kmiecik, D.; Gramza-Michałowska, A. The chemical composition and nutritional value of chia seeds—Current state of knowledge. *Nutrients* **2019**, *11*, 1242. [CrossRef] [PubMed]
55. Ramadan, M.F. Nutritional value, functional properties and nutraceutical applications of black cumin (Nigella sativa L.): An overview. *Int. J. Food Sci. Technol.* **2007**, *42*, 1208–1218. [CrossRef]
56. Bovell-Benjamin, A.C. Sweet potato: A review of its past, present, and future role in human nutrition. *Adv. Food Nutr. Res.* **2007**, *52*, 1–59. [CrossRef] [PubMed]
57. Zubr, J. Oil-seed crop: Camelina sativa. *Ind. Crop. Prod.* **1997**, *6*, 113–119. [CrossRef]
58. Maghrebi, M.; Noori, R.; Bhattarai, R.; Mundher Yaseen, Z.; Tang, Q.; Al-Ansari, N.; Mehr, A.D.; Karbassi, A.; Omidvar, J.; Farnoush, H.; et al. Iran's Agriculture in the Anthropocene. *Earths Future* **2020**, *8*, e2020EF001547. [CrossRef]
59. Grote, U.; Fasse, A.; Nguyen, T.T.; Erenstein, O. Food security and the dynamics of wheat and maize value chains in Africa and Asia. *Front. Sustain. Food Syst.* **2021**, *4*, 617009. [CrossRef]
60. Bai, Y.; Costlow, L.; Ebel, A.; Laves, S.; Ueda, Y.; Volin, N.; Zamek, M.; Masters, W.A. Retail prices of nutritious food rose more in countries with higher COVID-19 case counts. *Nat. Food* **2022**, *3*, 325–330. [CrossRef]
61. Gruni, G. *The EU, World Trade Law and the Right to Food: Rethinking Free Trade Agreements with Developing Countries*, 1st ed.; Bloomsbury Publishing: London, UK, 2018.

62. Narciso, J.O.; Nyström, L. Breathing New Life to Ancient Crops: Promoting the Ancient Philippine Grain "Kabog Millet" as an Alternative to Rice. *Foods* **2020**, *9*, 1727. [CrossRef]
63. Kozicka, M.; Groot, J.C.; Gotor, E. Can crop diversity strengthen small-scale farmers' resilience? In *Agrobiodiversity Index Report 2019: Risk and Resilience*, 1st ed.; Bailey, A., Ed.; Bioversity International: Rome, Italy, 2019.
64. Hertel, T.; Elouafi, I.; Tanticharoen, M.; Ewert, F. Diversification for enhanced food systems resilience. *Nat. Food* **2021**, *2*, 832–834. [CrossRef]
65. Bellon, M.R.; Kotu, B.H.; Azzarri, C.; Caracciolo, F. To diversify or not to diversify, that is the question. Pursuing agricultural development for smallholder farmers in marginal areas of Ghana. *World Dev.* **2020**, *125*, 104682. [CrossRef]
66. Gil, J. Food trade and resilience. *Nat. Food* **2020**, *1*, 150. [CrossRef]
67. Rivera, W.M.; Qamar, M.K. Food security. In *Agricultural Extension, Rural Development and the Food Security Challenge*, 1st ed.; Food and Agriculture Organization of the United Nations: Rome, Italy, 2003.
68. Zanella, L.; Vianello, F. Functional food from endangered ecosystems: *Atriplex portulacoides* as a case study. *Foods* **2020**, *9*, 1533. [CrossRef] [PubMed]
69. Balakrishnan, G.; Schneider, R.G. The Role of Amaranth, Quinoa, and Millets for the Development of Healthy, Sustainable Food Products—A Concise Review. *Foods* **2022**, *11*, 2442. [CrossRef] [PubMed]
70. Nicholson, C.C.; Emery, B.F.; Niles, M.T. Global relationships between crop diversity and nutritional stability. *Nat. Commun.* **2021**, *12*, 5310. [CrossRef] [PubMed]
71. Brander, M.; Bernauer, T.; Huss, M. Improved on-farm storage reduces seasonal food insecurity of smallholder farmer households–Evidence from a randomized control trial in Tanzania. *Food Policy* **2021**, *98*, 101891. [CrossRef]
72. European Commission. Enhancing Agricultural Biodiversity. Available online: https://agriculture.ec.europa.eu/sustainability/environmental-sustainability/biodiversity_en (accessed on 14 October 2022).
73. The Environmental Impacts of Agricultural Intensification. Available online: https://iaes.cgiar.org//sites/default/files/pdf/Environmental%20Impacts%20of%20Ag%20Intensification%20TN9_July2020.pdf (accessed on 14 October 2022).
74. Mazac, R.; Meinilä, J.; Korkalo, L.; Järviö, N.; Jalava, M.; Tuomisto, H.L. Incorporation of novel foods in European diets can reduce global warming potential, water use and land use by over 80%. *Nat. Food* **2022**, *3*, 286–293. [CrossRef]
75. Angeli, V.; Miguel Silva, P.; Crispim Massuela, D.; Khan, M.W.; Hamar, A.; Khajehei, F.; Graeff-Hönninger, S.; Piatti, C. Quinoa (*Chenopodium quinoa* Willd.): An overview of the potentials of the "golden grain" and socio-economic and environmental aspects of its cultivation and marketization. *Foods* **2020**, *9*, 216. [CrossRef] [PubMed]
76. Andreotti, F.; Bazile, D.; Biaggi, C.; Callo-Concha, D.; Jacquet, J.; Jemal, O.M.; King, O.I.; Mbosso, C.; Padulosi, S.; Speelman, E.N.; et al. When neglected species gain global interest: Lessons learned from quinoa's boom and bust for teff and minor millet. *Glob. Food Secur.* **2022**, *32*, 100613. [CrossRef]
77. EIP-AGRI. Non-Chemical Weed Management in Arable Cropping Systems-Final Report. Available online: https://ec.europa.eu/eip/agriculture/sites/default/files/eip-agri_fg_non-chemical_weed_management_final_report_2020_en.pdf (accessed on 14 October 2022).
78. Lucchi, A.; Benelli, G. Towards Pesticide-Free Farming? Sharing Needs and Knowledge Promotes Integrated Pest Management. *Environ. Sci. Pollut. Res.* **2018**, *25*, 13439–13445. [CrossRef]
79. World Obesity Atlas 2022. Available online: https://www.worldobesityday.org/assets/downloads/World_Obesity_Atlas_2022_WEB.pdf (accessed on 14 October 2022).
80. WHO. European Regional Obesity Report 2022. Available online: https://apps.who.int/iris/bitstream/handle/10665/353747/9789289057738-eng.pdf (accessed on 14 October 2022).
81. CDC. Centers for Disease Control and Prevention. Available online: https://www.cdc.gov/obesity/data/adult.html (accessed on 14 October 2022).
82. Plourde, G.; Fournier-Ross, E.; Tessier-Grenier, H.; Mullie, L.A.; Chassé, M.; Carrier, F.M. Association between obesity and hospital mortality in critical COVID-19: A retrospective cohort study. *Int. J. Obes.* **2021**, *45*, 2617–2622. [CrossRef]
83. World Bank. Food Finance Architecture-Financing a Healthy, Equitable and Sustainable Food System. Available online: https://documents1.worldbank.org/curated/en/879401632342154766/pdf/Food-Finance-Architecture-Financing-a-Healthy-Equitable-and-Sustainable-Food-System.pdf (accessed on 14 October 2022).
84. Hutchins, S.H. Sustainable Agriculture in the US vs. the EU: A Comparative Look at Different Approaches to Similar Objectives. *CSA News* **2021**, *66*, 24–34. [CrossRef]
85. USDA. U.S. Agriculture Innovation Strategy: A Directional Vision for Research. Available online: https://www.usda.gov/sites/default/files/documents/AIS.508-01.06.2021.pdf (accessed on 14 October 2022).
86. African Union Commission Agenda 2063: The Africa We Want. Available online: https://www.afdb.org/fileadmin/uploads/afdb/Documents/Policy-Documents/Agenda2063_Popular_Version_English.pdf (accessed on 3 November 2022).
87. Asian Development Bank. Strategy 2030. Available online: https://www.adb.org/sites/default/files/institutional-document/435391/strategy-2030-brochure.pdf (accessed on 3 November 2022).

MDPI AG
Grosspeteranlage 5
4052 Basel
Switzerland
Tel.: +41 61 683 77 34

Foods Editorial Office
E-mail: foods@mdpi.com
www.mdpi.com/journal/foods

www.ingramcontent.com/pod-product-compliance
Lightning Source LLC
Chambersburg PA
CBHW062044220926
51850CB00010B/2660